T0145076

Data Acquisition for Sensor Systems

Sensor Physics and Technology Series
Series editors:
Professor K. T. V. Grattan
Centre for Measurement, Instrumentation and Applied Physics
City University
London, UK

Dr A. Augousti
School of Applied Physics
Kingston University
Kingston-upon-Thames, UK

The *Sensors Physics and Technology Series* aims to bring together in a single series the most important developments in the rapidly-changing area of sensor technology and applications. It will present a snapshot of the range of effort which is being invested internationally in the development of novel types of sensors. New workers in the area of sensor technology will also be catered for with an introduction to the subject through the provision of tutorial guides. Volumes may be sensor technology or applications oriented, and will present recent results from the cutting edge of research of a compact monograph format.

Topics covered will include:

- optical sensors: free-space sensors
- optical sensors: guided wave sensors
- solid-state sensors
- biosensors
- microwave sensors
- ultrasonic sensors
- process tomography
- control of networked sensors system control and data acquisition
- medical instrumentation
- infrared sensors
- chemical and biochemical sensing
- environmental sensing
- industrial applications

Titles available:

1. Biosensors
   Tran Minh Cahn
2. Fiber Optic Fluorescence Thermometry
   K. T. V. Grattan and Z. Y. Zhang
3. Silicon Sensors and Circuits
   F. Wolffenbuttel
4. Ultrasonic Measurements and Technologies
   Štefan Kočiš and Zdenko Figura
5. Data Acquisition for Sensor Systems
   H. Rosemary Taylor

Data Acquisition for Sensor Systems

H. ROSEMARY TAYLOR

formerly
Lecturer in Electrical Engineering and Electronics
at
UMIST
The University of Manchester
Institute of Science and Technology

CHAPMAN & HALL
London · Weinheim · New York · Tokyo · Melbourne · Madras

Published by Chapman & Hall, 2–6 Boundary Row, London SE1 8HN, UK

Chapman & Hall, 2–6 Boundary Row, London SE1 8HN, UK

Chapman & Hall GmbH, Pappelallee 3, 69469 Weinheim, Germany

Chapman & Hall USA, 115 Fifth Avenue, New York, NY 10003, USA

Chapman & Hall Japan, ITP-Japan, Kyowa Building, 3F, 2-2-1 Hirakawacho, Chiyoda-ku, Tokyo 102, Japan

Chapman & Hall Australia, 102 Dodds Street, South Melbourne, Victoria 3205, Australia

Chapman & Hall India, R. Seshadri, 32 Second Main Road, CIT East, Madras 600 035, India

---

First edition 1997

Typeset in 10/12pt Times by Florencetype Ltd, Stoodleigh, Devon

Printed in Great Britain by St Edmundsbury Press Ltd, Bury St Edmunds, Suffolk

ISBN 0 412 78560 9

A catalogue record for this book is available from the British Library

Library of Congress Catalog Card Number: 96–72032

∞ Printed on permanent acid-free text paper, manufactured in accordance with ANSI/NISO Z39.48-1992 and ANSI/NISO Z39.48-1984 (Permanence of Paper).

Contents

Preface

'Data acquisition' is concerned with taking one or more analogue signals and converting them to digital form with sufficient accuracy and speed to be ready for processing by a computer. The increasing use of computers makes this an expanding field, and it is important that the conversion process is done correctly because information lost at this stage can never be regained, no matter how good the computation. The old saying – garbage in, garbage out – is very relevant to data acquisition, and so every part of the book contains a discussion of errors: where do they come from, how large are they, and what can be done to reduce them?

The book aims to treat the data acquisition process in depth with less detailed chapters on the fundamental principles of measurement, sensors and signal conditioning. There is also a chapter on software packages, which are becoming increasingly popular. This is such a rapidly changing topic that any review of available programs is bound to be out of date before the book reaches the readers. For this reason, I have described the data handling which is available in various types of program and left it to the reader to select from whatever is on the market at the time.

Scientists and engineers of any discipline who wish to use data acquisition in their work will find information here to enable them to make an informed selection from the many systems of hardware and software which are now on the market, and to optimize the parameters of the system for each application. The book may also be used by circuit engineers designing electronic instruments which need to incorporate some data acquisition for a specialized purpose.

It is in the nature of the subject that a topic may be confusing, or even incomprehensible, because some relatively minor point of jargon or protocol has not been explained. It has therefore been necessary to include some material which most readers will have met before and will consider quite elementary, and this is the reason why the academic level is not uniform.

The chapters are presented in the order in which a signal passes through the system. Over many years of research and development work on measuring instruments for applications in a variety of industries, I have found it necessary to study topics which are normally classified as parts of physics, metrology, computation or mechanical engineering as well as electrical engineering. The most useful of these topics as applied to data acquisition are presented in Chapters 2, 3, 11 and 12.

The core of the book, Chapters 4–10, concerns the devices used in a data acquisition system and how they are put together. These have their origins in lectures on Data Acquisition given by the author at UMIST, most of which formed part of M.Sc. courses in 'Instrument Design and Application', and 'Communications and Digital Electronics'. Students on these courses came from a wide variety of backgrounds and countries. They were mostly, but by no means exclusively, graduates in electrical or electronic engineering.

At the end of most chapters there are application examples with fully worked solutions. Most of these examples were originally set in examinations at UMIST and many of them are based on practical experience.

I have tried to minimize the use of abbreviations and to define each abbreviation the first time it occurs. However, a few terms are so widely used that writing them in full every time would make the text too cumbersome. So, operational amplifiers become op-amps and least significant bits become LSBs, and so on.

The problem of designing a data acquisition system, or indeed any measuring system, is finding the optimum accuracy, sensitivity, speed and reliability for a particular application at the minimum cost.

The ready availability of small computers, plug-in acquisition boards and powerful versatile software means that data acquisition is being used by more and more people for more varied tasks each year. However good the software, the results can never be more accurate than the incoming data, which is only another way of saying that the strength of a chain is in the weakest link. I have tried to demonstrate here that data acquisition is not a black art, but a fusion of parts of physics, electrical engineering, computer science and mathematics into a powerful tool. If this book has helped you to understand how these systems work and to improve the accuracy of your data or to avoid unnecessary expense, then it has served its purpose.

Rosemary Taylor, 1996

Acknowledgements

I should like to acknowledge the help of my husband and of my friends and former colleagues in the departments of Electrical Engineering at UMIST and at the University of Manchester, particularly those who taught on the Instrument Design and Application course. Thanks are also due to the many other people, from students to external examiners, who have made constructive comments on the lectures and examination questions on which this book is based.

I am grateful to the following organizations for permission to reproduce or adapt their copyright material:

- The Registrar of UMIST for examination questions.
- Solartron, for details of the pulse width converter in Section 9.7 and Figure 9.11. This converter is the subject of UK patents 1273790 and 1434414.
- AVX Corporation for the Capacitor Dielectric Comparison Chart, Appendix C and Table 9.1.
- Tektronix (UK) Ltd for the ASCII and GPIB code chart, Table 11.5.
- Analog Devices Inc. for part of the data sheet for AD 582, adapted for Figure 5.4.
- BSI Standards for definitions quoted in Section 2.2. (Extracts from BS 5233:1986 are reproduced with the permission of BSI. Complete copies can be obtained by post from BSI Sales, 389 Chiswick High Road, London, W4 4AL.)
- The IEEE for Figures 11.1 and 11.2 which are adapted from IEEE Std.488.1-1987, *IEEE Standard Digital Interface for Programmable Instrumentation (Reaffirmed 1994)*, copyright © 1994 by the Institute of Electrical and Electronic Engineers Inc. The IEEE takes no responsibility for and will assume no

liability for damages resulting from the reader's misinterpretation of said information resulting from the placement and context in this publication. Information is reproduced with permission of the IEEE.

Information gleaned from the manufacturers' catalogues is acknowledged in the text.

CHAPTER 1

Introduction

1.1 What is data acquisition?

There are many situations where data is required in digital form from a number of points simultaneously, either for the record or as a prerequisite to control. Not so very long ago, information was usually presented in analogue form by the position of a pointer on a dial, and an operator had to read a number of dials in order to control a machine or a process. As systems became faster and more complex, the operator's task became more difficult and, in some systems, impossible. Similarly, if a permanent record was required, a multichannel chart recorder was used. Even though most sensors still present their output as a voltage or current **analogue** of the quantity being measured (the **measurand**), it is now usual to convert the incoming analogue signal into **digital** form, i.e. into a series of numbers known as **data**. This change has been caused by the cost of employing an operator increasing at the same time as the cost of digital electronics and computers have been decreasing. In addition, a computer can read more information and act on it more quickly, to achieve **real-time control**.

Data acquisition is the recognized name for the branch of engineering dealing with collecting information from a number of analogue sources and converting it to digital form suitable for transmission to a computer, printer or alphanumeric display. A more descriptive, but less exact, title for this book would be 'Analogue and digital circuits for modern measuring instruments'.

The chapters which follow consider first the general principles of data acquisition systems. Then each unit of the system is considered in detail, including a critical description of currently available devices and techniques with a discussion of the errors introduced. Finally, we consider how these units may be combined together

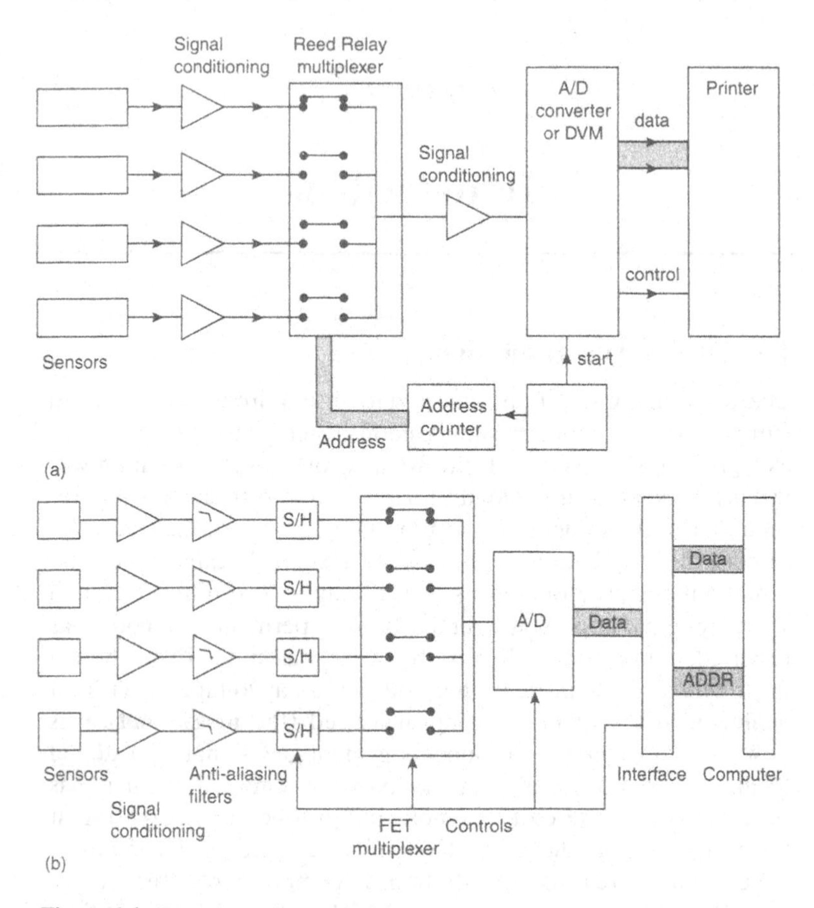

Figure 1.1 *Data acquisition systems: (a) low speed (up to 10 samples per second); (b) high speed.*

to form systems for specific purposes, and how the data may be transferred to a computer, displayed, analysed and recorded.

Figure 1.1(a) is typical of systems taking fewer than ten samples per second. Such systems are widely used in the process industries and may have several hundred input channels. Figure 1.1(b) represents a higher speed system, such as those used within instruments like storage oscilloscopes, and in digital audio and video systems.

1.2 Elements of a data acquisition system

The following elements are commonly found in data acquisition systems:

- **Sensors** or transducers are devices which convert the quantity which is to be measured (**measurand**) into a proportional electrical signal. This signal is an analogue of the measurand. It is debatable whether sensors should be included in a book on data acquisition systems. However, bearing in mind that 'garbage in = garbage out', a general overview and a discussion of the underlying principles should be useful, and is included in Chapter 3.
- **Signal conditioning** includes all the devices which convert the signal from the sensor to the correct level for the A/D (analogue to digital converter). It includes amplifiers, filters, and a.c. to d.c. converters. These are discussed in Chapter 4.
- **Anti-aliasing filters** are used to remove high-frequency signals which cannot be accurately converted. Aliasing is explained in Chapter 10.
- A **sample and hold** circuit is used before the multiplexer to sample several signal channels simultaneously or before the A/D to prevent the input to it changing while conversion is taking place. These applications are illustrated in Figure 1.1(b), and are described in Chapter 5 together with other applications of the circuit.
- The **multiplexer** (MUX) is a selector switch connecting one channel at a time to the A/D. It may be electromechanical as in Figure 1.1(a) or completely electronic as in Figure 1.1(b). Multiplexers are described in Chapter 6.
- The **analogue to digital converter** (A/D) is the heart of the system. Many designs exist and it is important to choose one which is suited to the application. For this reason, they are described in detail. Chapter 7 describes the two essential ingredients, a stable reference voltage and a code. As some A/D converters include a D/A converter these are described next, in Chapter 8, and then Chapter 9 describes a wide range of A/D converters and discusses the suitability of different types for various applications.
- The **clock** is the source of all the timing pulses needed to operate the units in the correct sequence. Figure 1.1(a) is a

simple system using an independent clock, but in Figure 1.1(b) control is by the computer clock. Instead of taking samples at regular intervals, it is straightforward to take a sample (or samples) every time a certain event occurs, such as every time a product comes off the production line.

- **Random access memory** (RAM) is used for the temporary storage of data. The system may utilize the RAM in the computer or RAM may be built into the acquisition system to take data from the converter, and hold it until the printer or computer is ready for it.
- A **digital to analogue converter** (D/A) may be driven by the computer if an analogue output is required as in a control system.
- An **interface** is a circuit to connect one element to another, and in particular to connect an A/D or a D/A to a microprocessor bus.
- A **bus** comprises a number of wires carrying related signals and each connected to several devices. The busses in Figure 1.1 are indicated by shading. As a bus links several units of a system, some standardization is necessary. Standard bus systems for various purposes are described in Chapter 9.

1.3 Complete data acquisition systems

Data acquisition systems may be assembled from the basic units, as described in Chapters 3–9, or they may be purchased in units of varying complexity.

1.3.1 Integrated circuits

Except for very specialized applications, analogue to digital converters, sample and hold circuits and solid state multiplexers, are always bought as integrated circuits. Prices depend on accuracy and speed and range from a few pounds to almost £100.

Integrated circuits which combine a number of functions are becoming increasingly common, for example:

- dual and quad converters;
- sampling converter (A/D + S/H);
- converter with instrumentation amplifier;
- converter with voltage reference on the same chip;

- converter with display drivers;
- converter with multiplexer;
- converter with multiplexer and S/H – complete data acquisition system;
- converter with serial output;
- converter with microprocessor and memory.

Most of these devices are available from more than one manufacturer with small variations in the type number. Thus AD574, ADC574 and HI1574 are all versions of the industry standard, 12-bit successive approximation A/D. Additional letters following the number indicate the package, the temperature range and accuracy grade of the device.

Microprocessors are also becoming increasingly common with a converter and/or RAM on board.

1.3.2 Data acquisition boards

Boards are available to various specifications, usually 12- or 16-bit resolution and up to 16 channels. These cost from a few hundred pounds each to over £1000 including essential software, and provide a quick and economical method of making a PC into a measuring instrument. They are also available for all the standard bus systems described in Chapter 11, so that an instrument may be assembled in a powered rack by plugging a number of boards, not necessarily all from the same manufacturer, into a standard backplane.

Boards are also available including the signal conditioning for the most widely used sensors – strain gauges, thermocouples and resistance thermometers.

1.3.3 Software for data acquisition

Software packages may be bought to control data acquisition boards and to analyse the resulting data. They are often intended for use with particular hardware. Chapter 12 explains the functions which are available. Useful reviews of such packages, including costings, are given by Johnson (1994), Barton (1991 and 1992), Taylor (1995) and Wright (1995).

1.3.4 Data loggers for process control

Complete systems including computer and software may be bought from the major instrument manufacturers, such as Solartron, Hewlett-Packard, Keithley and Fluke/Philips. The price obviously depends on the complexity of the system, but is of the order of several thousand pounds. These systems have a main unit with the digital voltmeter (DVM), computer and display. The input units are modular so that each system can be configured for the required number of channels, sensitivity and speed.

1.3.5 Digital voltmeters and multimeters

A **digital voltmeter** (DVM) comprises attenuators and amplifiers for various ranges, an A/D converter and a display. **A digital multimeter** (DMM) also includes circuits for measuring alternating voltage, direct and alternating current and resistance. Many DMMs also include some computation, memory and a bus interface to transmit the measurement to a computer or recorder. As people prefer reading numbers in decimal format to binary, the displayed measurement is always decimal and the number of digits indicates the resolution of the instrument. A 'half digit' can display only 0 or 1, so that a 3½-digit instrument has a maximum reading of ± 1999.

The more expensive instruments, i.e. 4½ digits and upwards, have some computational power and some include memory so that they can also be used as data loggers. A high quality data logger can be assembled by linking a scanner and a DVM to a computer by the IEEE-488 bus. Digital voltmeters intended to be used this way are described as **systems** DVMs, and allow the user some control over the measuring speed. There is usually a trade-off between speed and resolution. Simpler DVMs take about three measurements per second because they are intended mainly for visual reading. Digital voltmeters that also have capabilities for measuring current and resistance are more accurately designated digital multimeters (DMMs).

1.3.6 Choosing a system

With the help of the information in this book, it should be possible to choose the most appropriate system for a particular application. However when comparing prices, remember the hidden costs.

The integrated circuits (ICs) are cheap, but finding the right ones and designing boards takes engineers' time. So does writing software. Also, commercial systems have maintenance manuals, even if they have to be bought at extra cost, and if the system goes wrong it can be sent back to the manufacturer. Home-made systems are not always as well documented as they should be. Nevertheless, understanding exactly what is required and what is available should prevent you being mesmerized by overenthusiastic salesmen.

CHAPTER 2

Principles of measurement

2.1 Reasons for measuring

Measurements may be made for a number of reasons including:

- Fair trade – this is the oldest reason for measuring. In the 8th century BC, the prophet Micah wrote 'Can I overlook the infamous false measure, the accursed short bushel? Can I connive at false scales or a bag of light weights?' Micah Ch. 6, v. 10–11 (*New English Bible*).
- To make the components in a system compatible. For example, a door must fit in its frame, or a nut must screw on a bolt. Similarly, the supply voltage to an IC must not be large enough to cause damage or so small that the IC will not work. Accurate and reliable measurements were the precursors of mass production in both electrical and mechanical engineering.
- To permit control or to make the conditions correct. Many processes work best at a particular temperature, so the temperature has to be measured to control the heat input and maintain the optimum temperature.

Although measurements have been made for centuries, they are now required to be faster, more accurate and more reliable than ever before, and made without a human operator. This is the reason for the development of data acquisition systems.

2.2 What is measurement?

2.2.1 Definitions and terminology

Measurement is the process of comparing an unknown quantity with a standard of the same quantity, as in measuring length, or with standards of two or more related quantities, as in measuring velocity which has to be compared with both distance and time.

The comparison gives us an answer in terms of a standard multiplied by a pure number, and the units of the unknown are the same as the units of the standard. **Metrology** is the science of measurement and the **measurand** is the quantity being measured.

Many words which occur in normal English usage have much more precisely defined meanings when used in a measurement context. They were all defined and explained in British Standard BS 5233:1986 *Glossary of terms used in metrology*. This has been superseded by the BS *Vocabulary of Metrology* PD6461:Part 1:1995. A few of the most common are as follows:

2.2.2 Errors and the quality of a measurement

Several terms relate to measurement quality. **Accuracy** is the closeness of the measured value to the true value, and it is expressed in terms of the 'uncertainty'. **Uncertainty** is an estimate of the range of values within which the true value lies. Thus if a volt-

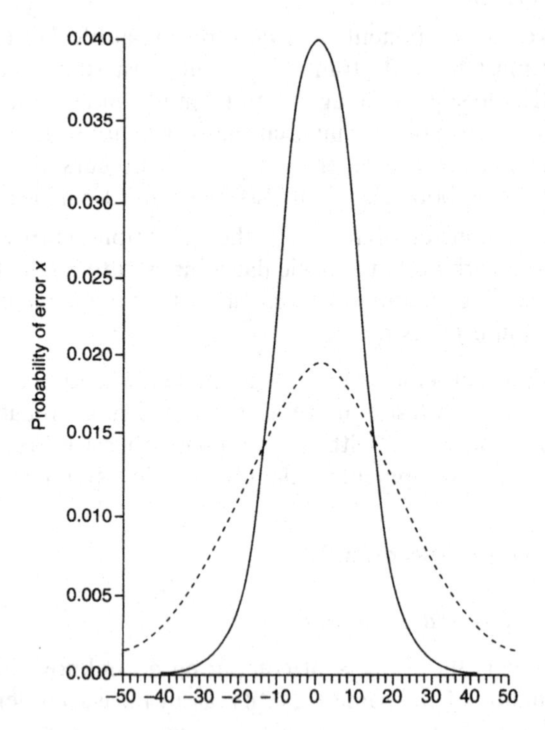

Figure 2.1 *Normal distribution of errors: (—) standard deviation = 10; (- - -) standard deviation = 20.*

meter is accurate to 1% and reads 10 V, we know that the true voltage is between 9.9 V and 10.1 V.

All measurements are subject to both **random errors** and **systematic errors**. The probability distribution of random errors and of the total error will be normal or Gaussian, as illustrated in Figure 2.1.

Precision is the closeness together of the measured values when the measurement is repeated. Examples of measurements which are precise but not accurate are a tape-measure which has stretched, or an A/D converter with an error in the value of its reference voltage. These are systematic errors.

BS 5233 distinguishes between repeatability and reproducibility. **Repeatability** is defined as the closeness of the measured values when repeated with the same method, observer, instrument, location and conditions. **Reproducibility** is defined as the closeness of the measured values when repeated with different methods, observers, instruments, locations, conditions and time. Thus repeatability is a measure of random error, but ignores systematic errors, whereas reproducibility takes account of both.

The **standard deviation** provides a measure of the width of the random error distribution curve, and has the symbol σ. The value σ^2 is the **variance** and is defined as

$$\sigma^2 = \sum_{i=1}^{i=N} \frac{(x_i - x_m)^2}{N - 1} \tag{2.1}$$

where N is the number of readings: x_1 to x_N and x_m is the mean value of the readings. In this definition

50% of the readings will be within $\pm\, 0.6745\sigma$ of the mean
68.3% will be within $\pm\, \sigma$ of the mean
95.5% will be within $\pm\, 2\sigma$ of the mean
99.7% will be within $\pm\, 3\sigma$ of the mean

To specify the uncertainty in full, we should also state not only the error limits, but the probability that the correct value lies within those limits. These are the **confidence levels**. The theory of errors is treated in every elementary book on statistics, but one that is particularly suited to practising engineers is Holman (1993). The topic is also summarized in Kaye (1986).

Digital instruments and data acquisition systems have the additional uncertainty of quantization error. In a binary system this is ± ½ a least significant bit (LSB), and in a decimal system ± ½ a least significant digit. The accuracy of a digital meter is specified in the form '$\pm x\%$ of reading $\pm y\%$ of full-scale' and $y\%$ of full-scale is always at least one LSD. The probability distribution for quantization error is uniform from $-\frac{1}{2}$ LSB to $+\frac{1}{2}$ LSB and zero otherwise. More details are give in Chapter 10.

The **resolution**, or **discrimination**, is the smallest change in the measurand that can be detected. In an analogue system, the resolution is limited by the noise level, but in a digital system it is limited by the size of the least significant bit or the noise level, whichever is larger.

Sensitivity is the relationship of the change in response to the change in stimulus. An example of instruments whose accuracy is too small for their sensitivity and resolution can be seen in any shop which sells cheap liquid-in-glass thermometers. They can produce a noticeable change in reading for each degree change in temperature, but the readings of a boxful of thermometers commonly vary by several degrees.

The terms introduced above are interrelated but they are **not** interchangeable. They are often confused, particularly in the nonscientific press.

2.3 Units and standards

Units are defined theoretically, and **standards** are their physical representations. The experiment to measure a standard in terms of its unit is called **realizing** the unit. As experimental accuracy has increased, it has become necessary to improve the realizations from time to time and occasionally to revise the unit definition. Some recent changes are described below.

In this context, a standard denotes the highest possible accuracy, not as in the everyday English usage of standard, as in the standard and de-luxe models of, say, a car. The word standard is also used in English for the documents such as British or IEEE standards. The French word *etalon* is sometimes used for physical standards to avoid this confusion (BS 5233:1986).

A good general review of units and standards was given by Bailey (1982), but several of the definitions he gives have changed since (see below). A compact summary of the current values of the SI units and the experiments to realize them is published as a wallchart by the UK National Physical Laboratory (1989).

2.4 Système Internationale d'Unités

Local units may have served isolated communities, but two communities who wish to trade must have some common units of measurement, and to cut a long and fascinating story short, the Système Internationale d'Unités (SI) was developed from the metric system, accepted by international agreement and it now serves the scientific and engineering communities of all nations.

To cover all scientific measurements SI needs seven units:

- metre – length
- kilogram – mass
- second – time
- ampere – electric current
- candela – luminous intensity
- kelvin – temperature
- mole – amount of a substance.

All other units can be derived from these seven, and no conversion factors are involved. For example:

- 1 m s^{-2} is 1 unit of acceleration, which has no special name;
- 1 newton (N) is the force to produce an acceleration of 1 m s^{-2} in a mass of 1 kg, therefore 1 N = 1 kg m s^{-2};
- 1 newton moved through 1 metre dissipates 1 joule (J) of energy, therefore 1 J = 1 kg m^2 s^{-2};
- a power of 1 watt (W) uses energy at the rate of 1 joule per second, therefore 1 W = 1 kg m^2 s^{-3};
- a current of 1 amp flowing between points at a potential difference of 1 volt also dissipates 1 watt, therefore 1 volt = 1 kg m^2 s^{-3} A^{-1}.

2.5 Standards

Standards may be either **material** or **recipe**. Originally all were material, such as the Standard Pound and the Imperial Standard Yard. Material standards have several major disadvantages:

- If each nation has its own standards, how do we know whether they are the same? If not, by how much do they differ?
- Material standards may alter with time or may be destroyed as were the Standard Pound and the Imperial Standard Yard when the Houses of Parliament were burnt down in 1834.

Table 2.1 Standards organizations

Common abbreviation	*Organization*
International organizations	
CGPM	Conférence Général des Poids et Measures – has responsibility for SI at political level; meets every few years
CIPM	Comité International des Poids et Measures – committee of scientists carrying out decisions of CGPM and supervising BIPM
BIPM	Bureau International des Poids et Measures – laboratory at Sèvres near Paris, set up in 1875; responsible for development, maintenance and dissemination of BIPM standards
CCE	Comité Consultatif d'Electricité – Subcommittee of CIPM responsible for electricity
National standards laboratories	
NPL	National Physical Laboratory, Teddington, Middlesex, UK
NIST	National Institute of Standards and Technology, Washington, DC, USA
NRC	National Research Council, Ottowa, Canada
CSIRO	Commonwealth Scientific and Industrial Research Organisation, Sydney, Australia
PTB	Physikalisch Technische Bundesanstalt, Braunschweig and Berlin, Germany
Publishers of documentary standards	
BSI	British Standards Institution, London
ISO	International Standards Organisation
IEEE	Institute of Electrical and Electronic Engineers, New York, USA
IEC	International Electrotechnical Commission
EIA	Electronic Industries Association
CCITT	International Telegraph and Telephone Consultative Committee

- There can be only one material standard of each unit, but recipe standards can be realized by anyone with the necessary knowledge and equipment. In practice this means the national standards laboratories of all industrial countries. These include the NPL in Britain, the NIST in the USA, and the PTB in Germany.

Only the kilogram is still defined by a material standard; all other units use recipe standards. Table 2.1 lists some of the organizations involved in setting up and maintaining the international measurement systems.

2.6 Unit and standard of mass

The unit is the **kilogram** and the standard is a cylinder made of platinum–iridium in 1880. It is kept at BIPM at Sèvres under controlled conditions. Each national standards laboratory has a copy which is occasionally compared with it.

2.7 Unit and standard of time

As everyone knows, the standard of time used to be based on the rotation of the earth. The **second** is now defined as 'the duration of 9 192 631 770 periods of the radiation corresponding to the transition between the two hyperfine levels of the groundstate of the caesium-133 atom'. This definition is implemented by the atomic clock and every national standards laboratory has one. No longer is there any need for comparisons. The BBC 16 kHz transmissions from Rugby with seconds markers and coded indications of the time of day are derived from the atomic clock at Teddington. Mazda (1987) lists time signal transmissions all over the world. Time can be measured more accurately than any other quantity. The overall accuracy of the broadcasts is about 1 in 10^{11}, although the accuracy of the caesium clock itself is higher, at about 1 in 10^{13}.

2.8 Unit and standard of length

The unit is the **metre**. The standard was originally intended to be a recipe standard equal to the length of the polar quadrant passing through Paris times 10^{-7}. This was in practice replaced by a material standard kept at Sèvres until 1960 when it was replaced by a recipe standard based on radiation emitted by krypton atoms. However when lasers were developed which were more stable than the krypton lamp, a new standard had to be found. Originally there had been independent standards for the second and the metre, and the velocity of light c, which is a fundamental constant, had been measured experimentally. In 1983 the CGPM changed the definition of the metre to 'the length of the path travelled by light during the time interval of 1/299 792 458 s'. At the time, 299 792 458 m s^{-1} was the best experimental value of the velocity of light and all other electromagnetic radiation, so the metre is now a defined constant, but has the same value as before.

At the NPL the metre is realized by an iodine-stabilized helium–neon laser. This has been accurately related to the new definition and is reproducible to three parts in 10^{11}.

A popular account of the history of length standards and the reason for the redefinition was given by Wilkie (1983).

2.9 Electrical units and standards

2.9.1 ϵ_0 and μ_0

Electrical units are linked to mechanical units by the electrostatic and electromagnetic forces. The equations relating charge to electrostatic force include the constant ϵ_0, known as the **permittivity of free space**, and the magnetic equations include μ_0, the **permeability of free space**. ϵ_0 and μ_0 are themselves related by the equation

$$\epsilon_0 \times \mu_0 = c^2 \qquad (2.2)$$

where c is the velocity of electromagnetic radiation. Thus, to link electrical units to the mechanical ones defined above, and to keep a coherent system, SI has to define ϵ_0 or μ_0, but not both. The choice was to define μ_0 as $4\pi \times 10^{-7}$ H m^{-1} (henrys per metre). It follows that $\epsilon_0 = 8.845\,187\,817 \times 10^{-12}$ F m^{-1} (farads per metre).

2.9.2 Unit of electric current

Since 1948 the **ampere** or **amp** has been defined as 'that current, which if maintained in two straight parallel conductors of infinite length and negligible cross-section one metre apart in a vacuum would produce a force of 2×10^{-7} N between the conductors'. Although not obvious at first sight, this is exactly equivalent to the definition above that $\mu_0 = 4\pi \times 10^{-7}$ H m^{-1}.

The experiment to measure a current in terms of the SI amp is called realizing the amp, and for many years it was done with great difficulty using a current balance. This has been superseded by Kibble's moving coil experiment which realizes the watt. From this experiment, the amp is known to 8 parts in 10^8, and the volt to 2.5 in 10^8. It is interesting to note that 8½-digit voltmeters are commercially available from several manufacturers, and 8½ digits is a resolution of 1 in 10^8.

2.9.3 Units of potential and impedance

It is not possible to maintain a 'standard amp' so national standards laboratories maintain standards of the **volt** and the **ohm**. Several important new experiments that have been developed in the last few decades are:

- the Kibble moving coil experiment – the realization of the watt;
- the Josephson junction – the maintained standard of the volt;
- the Thompson–Lampard calculable capacitor – the realization of the ohm;
- the von Klitzing quantized Hall device which is the maintained standard of resistance.

It is beyond the scope of this book to describe these experiments, but Kibble (1986) gave a good general description. More detailed descriptions will be found in Kibble *et al.* (1988), Hartland (1988), Thompson (1968) and Cutkosky (1974). After the new methods had been in use for a few years, it was found necessary to make slight changes to the values. The NPL leaflet *Changes in the values of the UK reference standards of electromotive force and resistance* (R.G. Jones, 1989), and the papers presented at the IEE colloquium on 'Changes in the values in the UK national reference standards for the volt and the ohm' (O.C. Jones, 1989) give details. The new values were implemented on 1 January 1990.

2.10 Units and standards of temperature

Temperature is defined from thermodynamic theory by the scale of a perfect gas thermometer which follows the equation

$$PV = RT \tag{2.3}$$

The constant volume gas thermometer provides a nearly perfect scale, but is a very difficult instrument to use and is quite unsuited to practical applications. Over the years there have been a number of definitions of a practical scale and the current one is **ITS-90** (Rusby, 1989; Preston-Thomas, 1989). This scale came into use on 1 January 1990 at the same time as the revised standards for the ohm and the volt. ITS-90 defines the unit of temperature, the **kelvin** (not degree kelvin) as '1/273.16 of the thermodynamic temperature of the triple point of water'.

The **triple point**, 273.16 K, is the temperature at which solid, liquid and gas coexist in equilibrium. ITS-90 uses the triple point because it can be reproduced much more reliably than the melting point of ice.

ITS-90 is defined by 16 fixed temperatures which can be reliably and accurately maintained in a laboratory. These points range from the triple point of hydrogen at 13.8033 K (–259.3467 °C) to the freezing point of copper at 1357.77 K (1084.62 °C). For the highest accuracy for calibrating other thermometers, temperatures between the fixed points are measured by specified instruments with specified interpolation equations. For most of the range 13.8 to 1234.93 K (the freezing point of silver, 961.78 °C), the standard instrument is the platinum resistance thermometer, and for temperatures above that a radiation thermometer is used.

The **degree celsius** is the same magnitude as the kelvin, but is measured from the melting point of ice which is 0.01° lower, i.e. 273.15 K.

Thus the temperature in °C = temperature in kelvin –273.15.

2.11 Traceability and calibration

If the data from a system is to be reliable, it should be possible to trace its calibration through an unbroken chain of comparisons, all having stated uncertainties back to the national or international standard. This process is called **traceability** and the chain is illustrated in Table 2.2.

In the UK, traceability used to be organized by the National Measurement and Accreditation Service (NAMAS) which was based at the NPL. In 1995, NAMAS merged with the National Accreditation Council for Certification Bodies to form the United Kingdom Accreditation Service (UKAS; Cronin (1995a,b)). An instrument manufacturer or a calibration organization can apply for accreditation for one or more parameters over specified ranges. Assessors will visit the laboratory and examine, among other things:

Table 2.2 Traceability

Calibration	*Where assessed*
Theoretical definition of unit ↓	CGPM
Realization experiment ↓	some standards laboratories
National standard ↓	national standards laboratories
Transfer standard ↓	NAMAS
Laboratory or reference standard ↓	NAMAS accredited laboratory
Working standard ↓	manufacturer's laboratory
Product	production line

- the technical knowledge of the staff;
- the test or calibration method(s);
- the appropriateness of the facilities, equipment and materials;
- that the equipment is regularly calibrated and maintained.

Each step in the traceability chain must be documented, and only certain staff at an accredited laboratory are authorized to sign test certificates. UKAS publish a directory listing the names and addresses of accredited laboratories. To remain accredited, the laboratory must be reinspected at regular intervals.

The calibration and accreditation procedures are tightly controlled by the following European standards (Rogers, 1995):

1. the ISO/IEC Guide 25 *General requirements for the competence of calibration and testing laboratories*, and its European equivalent EN 45001 *General criteria for the operation of testing laboratories*;
2. the ISO/IEC Guide 58 *Calibration and testing laboratory systems – General requirements for operation and recognition*, and its European equivalents EN 45002 *General criteria for the assessment of testing laboratories*, and EN 45003 *General criteria for laboratory assessment bodies*.

Accredited laboratories are allowed to issue calibration certificates bearing the NAMAS logo and the laboratory's number (Carter, 1995; Skinner, 1995).

CHAPTER 3

Sensors

3.1 Introduction

Sensors are the devices which generate the data for the system to acquire. There is an enormous variety on the market and it is very easy for a chapter on sensors to become a mere catalogue, and an incomplete one at that. The pages that follow explain the general principles and describe a few sensors as illustrations. Readers requiring more detail should refer to one of the many good books available which cover sensors in detail, such as Barney (1988); Doebelin (1990), Sydenham (1982, 1983), Sydenham *et al.* (1989), Bannister and Whitehead (1991) and Holman (1993).

3.2 Active and passive sensors

Sensors are classified as active and passive on the same basis as filters. A **passive sensor** has no power supply and all the energy it delivers to the next stage, the signal conditioning, is drawn from the measurand. Passive sensors are also known as self-generating sensors. For example, the early record player pick-ups drew all their acoustic energy from the kinetic energy of the record. Long-playing records with narrower grooves could not be developed until the mass of the pick-up had been reduced by several orders, with the introduction of amplification.

An **active sensor** is a modulator and can therefore deliver more energy to the next stage than it draws from the measurand. If the power supply is d.c., the output is modulated by the measurand, and has the same frequency. If the supply is a.c., the output is the carrier frequency with sidebands at ± signal frequency.

3.3 Effort and flow variables

Electrical, mechanical, acoustic and thermal energy are all measured in joules. Each form of energy has two associated variables, the **effort** variable and the **flow** variable (Table 3.1):

$$\text{effort} \times \text{flow} = \text{power}$$

$$\frac{\text{effort}}{\text{flow}} = \text{resistance (or impedance)}$$

For alternating variables which are not in phase the resistance is known as the impedance.

To avoid drawing energy from the measurand and reducing its value, sensors for an effort variable must have a high impedance and sensors for a flow variable must have a low impedance.

A sensor which is intrinsically sensitive to a flow variable may be converted to measure the corresponding effort variable by the addition of a high resistance. For example, a moving-coil movement may be converted to either a voltmeter by a series resistance or to an ammeter by a parallel (shunt) resistance. Similarly 'strain' is change in length divided by original length, and strain gauges are commonly used to measure force by being attached to

Table 3.1 *Effort and flow variables*

Effort	*Flow*	*Quantity*	*Energy*	*Power*	*Impedance*
Potential	current	charge	energy	power	impedance
v	i	$q = i \times t$	$v \times i \times t$	$v \times i$	$z = v/i$
Volt	amp	coulomb	joule	watt	ohm
Force	velocity	displace-ment	energy	power	impedance
F	dx/dt	x	$F \times x$	$F \times dx/dt$	$F/(dx/dt)$
newton	$m\ s^{-1}$	metre	joule	watt	$N\ s\ m^{-1}$
Temp-erature	heat flow	heat quantity	energy	power	thermal resistance
T	dQ/dt	Q	Q	dQ/dt	$R = T/(dQ/dt)$
kelvin	joule s^{-1}	joule	joule	watt	thermal ohm

an object of the appropriate stiffness. A moving-coil microphone relies on a thin diaphragm to convert acoustic pressure to displacement. There are many other examples of such two-stage converters.

Note that force may also be considered analogous to current, and which analogy is appropriate depends whether the components are arranged in series or in parallel. As heat is a form of energy, the quantity of heat is measured in the same units as energy, i.e. joules, although electrical charge and electrical energy have different units and are related via the voltage.

3.4 Static performance of sensors

Figure 3.1 illustrates the errors which can occur in sensors, and it will be noticed that this figure is very similar to Figure 3.5 for D/A converters.

3.4.1 Accuracy

The slope of the output versus input graph is the **sensitivity**. In a few cases this can be calculated, but normally the sensor is calibrated by applying a known value of the input and observing the output. If this test is done slowly, the result is known as the static calibration. Mass-produced sensors have a specification which quotes a value for sensitivity, based on small numbers of carefully tested samples, and a tolerance band. To be acceptable the sensitivity of a particular sensor must be within the tolerance. A sensor of this type is assumed to have the specified sensitivity and the tolerance is treated as the uncertainty which limits the accuracy. More expensive sensors are supplied with individual calibration certificates. In Figure 3.1 the thin line represents the correct or **nominal** sensitivity. Gain error can be compensated for in the software or in the signal conditioning, and on some amplifiers the user sets a dial to the sensor sensitivity as given in the calibration certificate, and then the combined sensitivity of the sensor and amplifier is standardized.

3.4.2 Offset

The offset is the output measured at zero input (or whatever input should give zero output). Offset can be compensated for in the signal conditioning or the software.

3.4.3 Linearity

The output of a sensor is a function of the input, but not necessarily proportional. An example of a very nonlinear sensor is a radiation type of temperature sensor. Thermocouples are approximately linear for small temperature changes, but are nonlinear over their full range. However, most other sensors are sufficiently close to linear that the deviation from linearity can be classed as an error. Figures 3.1(c)–(e) show the same curve with

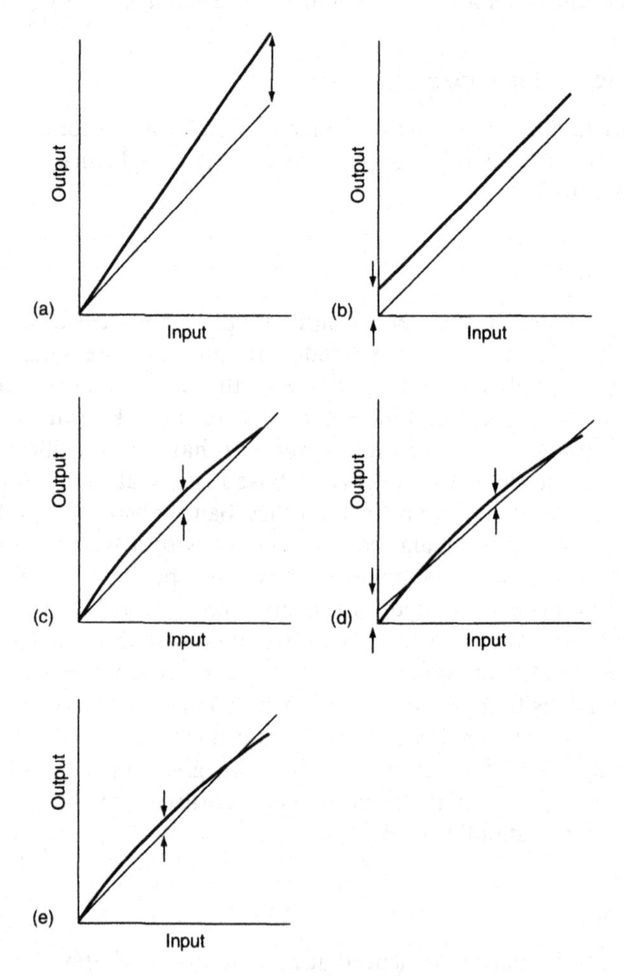

Figure 3.1 *Errors in sensors: (a) gain error, (b) offset, (c) terminal linearity error, (d) independent linearity error, (e) zero-based linearity error.*

the nonlinearity specified in three different ways. In (c) the actual output is compared with a straight line joining the two ends of the curve and the maximum difference between these two lines is the **terminal linearity error**. This is the definition which is also used for D/A and A/D converters. In (d) the output is compared with the best straight line through all the calibration points. The best straight line can be found by a least-squares fit computation. This gives the **independent linearity error**. This definition is often used for sensors and gives an error approximately one-third of the terminal linearity error for the same curve. When reading a linearity specification, check which method has been used. As in the example in (d), the best straight line may not pass through zero, even though there is really no offset. The third method illustrated in (e) uses the best straight line which passes through zero.

3.5 Dynamic performance of sensors

3.5.1 First- and second-order sensors

In addition to any errors which may be present when a sensor is calibrated slowly, there may be other errors which depend on how the input changes with time. These depend on whether the sensor includes any elements which can store energy.

If there is one element which can store energy, the system is first order, and if there are two it is second order. These are so common that it is worth studying the response in some detail.

Electrical energy comes in two forms: electrical, stored in capacitors, and magnetic, stored in inductors. Mechanical energy comes in two forms: potential, stored in springs, and kinetic, stored in moving masses. However, thermal energy comes in only one form. There is no thermal analogy to inductance.

3.5.2 Example of a first-order system: a thermometer

Consider a thermometer of thermal capacitance C, with a thermal resistance R, in the boundary layer to the bulk of the liquid.

The thermometer is originally at ambient temperature θ_0, and is suddenly plunged into liquid at temperature θ_F above ambient. The temperature of the thermometer at time t is θ above ambient. Heat flows through the thermal resistance at a rate

$$\frac{\theta_F - \theta}{R}$$

and the temperature of the thermometer rises at

$$\frac{d\theta}{dt} = \frac{\theta_F - \theta}{RC}$$

Therefore,

$$\int \frac{d\theta}{\theta_F - \theta} = \int \frac{dt}{RC} \tag{3.1}$$

and

$$\theta = \theta_F \times (1 - e^{t/RC}) \tag{3.2}$$

The term RC is the **time constant** of the system, and is the time taken for θ to reach 63.2% of θ_F (63.2% = $1-e^{-1}$).

3.5.3 Example of a second-order system: a thermometer in a sheath

If the thermometer in the example above is enclosed in an outer sheath for better protection, it becomes a second-order or two-pole system as represented by Figure 3.2(a). Here, θ_1 is the temperature of the outer fluid, θ_2 is the temperature of the well and θ_3 is the temperature of the sensor itself; R_1 is the thermal resistance from the sheath to the thermometer and R_2 is the thermal resistance of the sheath and boundary layer; C_1 is the thermal capacity of the thermometer and C_2 of the sheath and the boundary layer. The analysis of this system is considerably more complicated than for a first order system, and one method is given below (Taylor and Rihawi, 1993).

The thermal circuit in Figure 3.2(b) can be analysed by mesh analysis, as for an electrical one. A convenient way of doing the calculations is to use Laplace transforms which are explained in most engineering mathematics texts. Scott (1960) is a particularly good example. The Laplace variable is s, and tables are published to transform a function of time $f(t)$ into a Laplace variable $F(s)$, and back again. This is worth doing because the algebra in the Laplace domain is much simpler than in the time domain. Inspection of Figure 3.2(b) gives

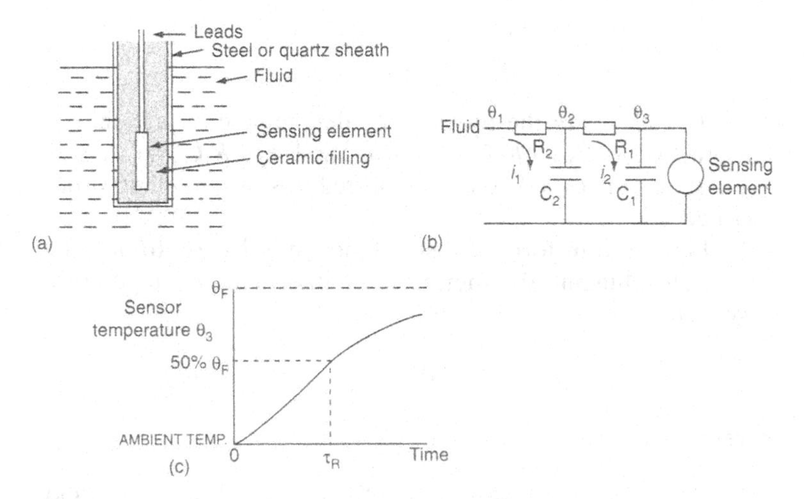

Figure 3.2 *Thermometer in a sheath: (a) section, (b) electrical analogue, (c) step function response.*

$$i_1(s) = \frac{\theta_1(s) - \theta_2(s)}{R_1} \tag{3.3}$$

$$i_2(s) = \frac{\theta_2(s) - \theta_3(s)}{R_2} \tag{3.4}$$

The mesh equations are

$$i_1(s)R_1 + \frac{i_1(s)}{sC_2} - \frac{i_2(s)}{sC_2} = \theta_1(s) \tag{3.5}$$

$$i_2(s)R_1 + \frac{i_2(s)}{sC_1} + \frac{i_2(s)}{sC_2} - \frac{i_1(s)}{sC_2} = 0 \tag{3.6}$$

Substituting for $i_1(s)$ and $i_2(s)$, and solving for θ_3 gives

$$\theta(s) = \frac{\theta_1(s)}{s^2R_1R_2C_1C_2 + s(R_1C_1 + R_2C_1 + R_2C_2) + 1}$$

$$= \frac{\theta_1(s)}{s^2T_1T_2 + s(T_1 + T_2) + 1}$$

$$= \frac{\theta_1(s)}{(sT_1 + 1)(sT_2 + 1)} \tag{3.7}$$

where T_1 and T_2 are the roots of the denominator equation.

In practice, $R_2C_2 >> R_1C_1$, $T_1 \approx R_2C_2$ and $T_2 \approx R_1C_1$, so that $T_1 >> T_2$. If R_1 or $C_2 = 0$, the system reduces to the first order described above.

The Laplace transform of a step function is $1/s$. So, to find the response to plunging the thermometer into fluid at temperature θ_F we write

$$\theta_1(s) = \frac{\theta_F}{s}$$

Therefore,

$$\theta_3(s) = \frac{\theta_F}{s(sT_1 + 1)(sT_2 + 1)} \tag{3.8}$$

Taking inverse Laplace transforms gives the response in the time domain:

$$\theta_3(t) = \theta_F \,(1 + B\,\mathrm{e}^{-t/T_1} + C\,\mathrm{e}^{-t/T_2})$$

A little more algebra gives the values of B and C in terms of T_1 and T_2:

$$\theta_3(t) = \theta_F\left(1 - \frac{T_1^{\,2}}{T_1 - T_2}\,\mathrm{e}^{-t/T_1} + \frac{T_2^{\,2}}{T_1 - T_2}\,\mathrm{e}^{-t/T_2}\right) \tag{3.9}$$

Figure 3.2(c) shows the shape of this function.

Most practical thermometers are second-order systems even if $T_1 >> T_2$, so BS 1904:1984 recommends that the response time should be defined as the time taken to reach 50% of θ_F, and that the 'time constant = 63.2% of θ_F' definition be reserved for first-order systems.

3.5.4 Example of second-order sensors: spring–mass systems

Mechanical and acoustic sensors all have some elements with mass, some with elasticity and some with damping, and the combination of these three determines the transient response and the frequency response of the sensor. The electrical analogy in this case includes inductance L, as well as capacitance C and resistance R. The measurand is a force or a pressure, but it is converted first to a displacement by the spring and second to an electrical parameter.

The force equation of a spring–mass system is

$$m \frac{\mathrm{d}^2 y}{\mathrm{d}t^2} + d \frac{\mathrm{d}y}{\mathrm{d}t} + ky = F \tag{3.10}$$

where y is displacement, F is force, m is mass, d is damping force and k is spring stiffness.

In Laplace notation this gives

$$y(s) = \frac{F(s)}{s^2 m + sd + k} \tag{3.11}$$

The poles of this function are given by the roots of the denominator, and the behaviour of the circuit depends on whether the roots are real, equal or complex. When $s = 0$, $y(0) = F(0)/k$, so the **static sensitivity** = $y(0)/F(0) = k^{-1}$ i.e., compliance.

Undamped response

When $d = 0$,

$$y(s) = \frac{F(s)}{s^2 m + k} \tag{3.12}$$

This is the equation of a system which, given the slightest disturbance, will oscillate at its **undamped natural frequency**:

$$f_0 = \frac{1}{2\pi} \left(\frac{k}{m} \right)^{1/2} \tag{3.13}$$

Therefore, for a good high-frequency response, k should be large, but for high sensitivity k should be small. Much of the skill in designing a sensor is in finding the best compromise and keeping the mass as small as possible, consistent with strength.

Damped response

Any practical sensor must have some damping, and there are several ways of quantifying damping. As above, d is the **damping force**, and d/m is the **damping factor**.

When the roots of the force equation are equal, $d^2 = 4\,km$, and damping is said to be **critical**.

The **damping ratio** (ζ) is the damping expressed as a fraction of critical damping:

$$\zeta = \frac{d}{2\sqrt{(mk)}} \tag{3.14}$$

Thus, for critical damping, $\zeta = 1$.

The equations of LCR circuits correspond exactly to those of mechanical spring–mass systems, but electrical engineers prefer to express damping in terms of the Q-factor, where $Q = (2\zeta)^{-1}$, or $\zeta = (2Q)^{-1}$. When damping is not negligible, the natural frequency is reduced to:

$$f_n = \frac{1}{2\pi}\left[\frac{k}{m}(1-\zeta^2)\right]^{1/2}$$
$$= f_0(1-\zeta^2)^{1/2} \tag{3.15}$$

The parameters m, d and k are useful to the sensor designer but the user is more interested in the performance parameters f_0, f_n and ζ, and the static sensitivity S_0:

$$S_0 = \frac{1}{k} \tag{3.16}$$

Now we may rewrite the force equation:

$$y(s)\left(\frac{s^2}{\omega_0{}^2} + \frac{2\zeta s}{\omega_0} + 1\right) = \frac{F(s)}{k} \tag{3.17}$$

Substituting any input function for $F(s)$ gives the corresponding function for $y(s)$. For a step function $F(s) = F/s$, where the step is a sudden rise from 0 to F, as in the example above for a thermometer. The step-function responses for underdamped ($\zeta = 0.05$ and $\zeta = 0.707$), critically damped ($\zeta = 1$) and overdamped ($\zeta = 10$) are shown in Figure 3.3.

For a periodic function, $F(t) = A \sin \omega t$, substitute $j\omega$ for s and $\beta\omega_0$ for ω:

$$y(\beta) = \frac{F(\beta)}{k(-\beta^2 + j2\zeta\beta + 1)} \tag{3.18}$$

This can be written in terms of amplitude and phase:

$$|y(\beta)| = \frac{F(\beta)}{k[(1-\beta^2)^2 + (2\zeta\beta)^2]^{1/2}} \tag{3.19}$$

$$\tan\phi = \frac{2\zeta\beta}{1-\beta^2} \tag{3.20}$$

From these equations we can see that, at low frequencies

$$y(\beta) = \frac{F(\beta)}{k}$$

and at high frequencies

$$y(\beta) = \frac{F(\beta)}{k\beta^2}$$

which decreases at 12 dB per octave.

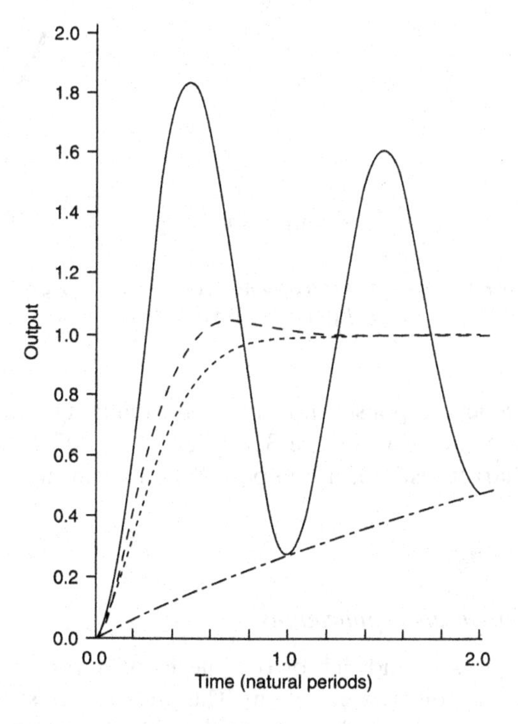

Figure 3.3 *Spring–mass system, step function response: (—) $\zeta = 0.05$, (- - -) $\zeta = 0.7071$, (· · · ·) $\zeta = 1.0$, (– · – ·) $\zeta = 10.0$.*

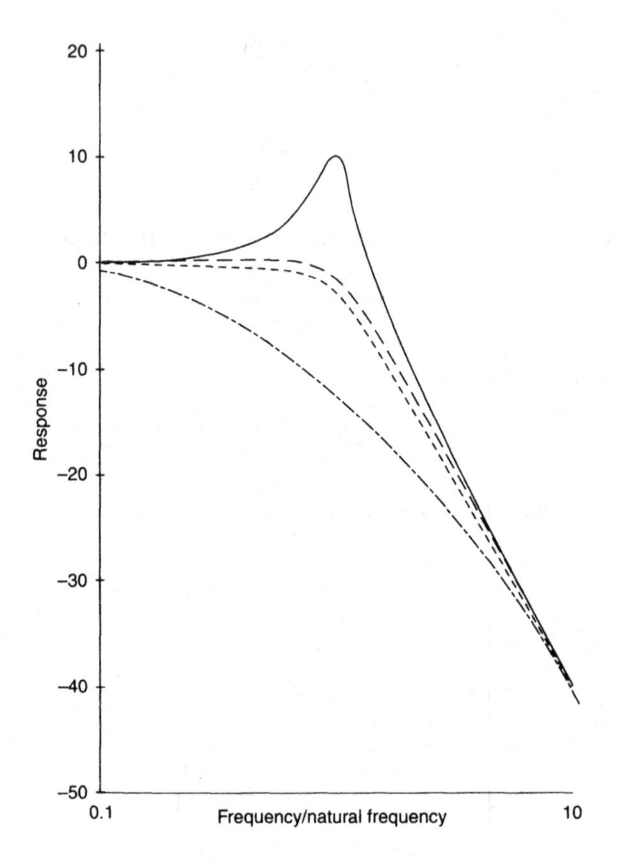

Figure 3.4 *Spring–mass system, frequency response: (——) $\zeta = 0.05$, (- - -) $\zeta = 0.7071$, (· · · ·) $\zeta = 1.0$, (– · – ·) $\zeta = 10.0$.*

The frequency responses for the same damping ratios as in Figure 3.3 are shown in Figure 3.4, where $\zeta = 0.707$ is the maximally flat characteristic of a two-pole Butterworth filter.

3.6 Strain gauges

3.6.1 Description and applications

A strain gauge is a sandwich of two pieces of thin plastic with a zig-zag of metal foil between them. The ends of the strip of foil are widened into pads which can be soldered to the external circuit; see Figures 3.5(a)–(c). A strain gauge is fixed to a piece of metal with a special adhesive so that if the metal is strained, the gauge is strained with it. Provided that the strain is in the sensitive direction, the strands of foil become longer and thinner and therefore their resistance increases. Since strain is proportional to stress (Hooke's law), a measurement of strain can provide a measurement of force and this is the normal use of strain gauges. Depending on the size and shape of the member to which the gauges are attached and their configuration, gauges may be used to measure (d) large and small forces in tension and compression, (e) pressure, (f) torsion and also (g) bending. A strain gauge sensor will be a second-order system with the limitations described above, because the gauges must be mounted on a member that has both elasticity and mass. The size of gauges ranges from a few millimetres to a few centimetres active length. Special patterns are available to measure pressure on a diaphragm and torsion (Figures 3.5(b) and (c)).

3.6.2 Strain

Strain is defined as

$$\frac{\text{extension}}{\text{original length}} \tag{3.21}$$

Strain is a dimensionless quantity and is generally a fraction which is much less than 1. The notation 1 μ strain, means that the extension is 10^{-6} times the original length. The range of strain which can be measured by resistance gauges is approximately from 10^{-5} (10 μ strain) to 10^{-2} (1%).

Stress is defined as

$$\frac{\text{force}}{\text{area}} \tag{3.22}$$

The units are the same as for pressure, i.e. pascals (N m^{-2}).

The **modulus of elasticity** is defined as

$$\frac{\text{stress}}{\text{strain}} \tag{3.23}$$

Again, the units are pascals. For linear extension and compression, this is Young's modulus which has the symbol E or Y.

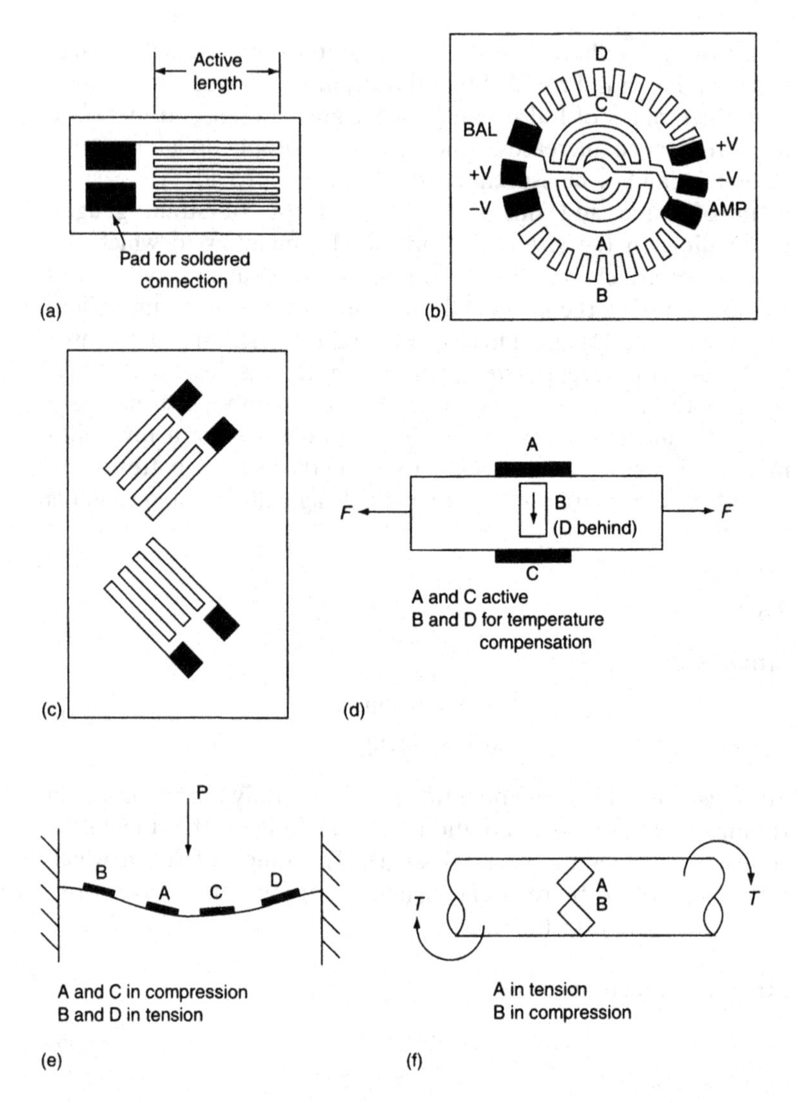

Figure 3.5 *Strain gauges: (a) gauge for tension or compression; (b) gauge for pressure on a diaphragm; (c) gauge for torsion; (d) application of (a); (e) application of (b); (f) application of (c).*

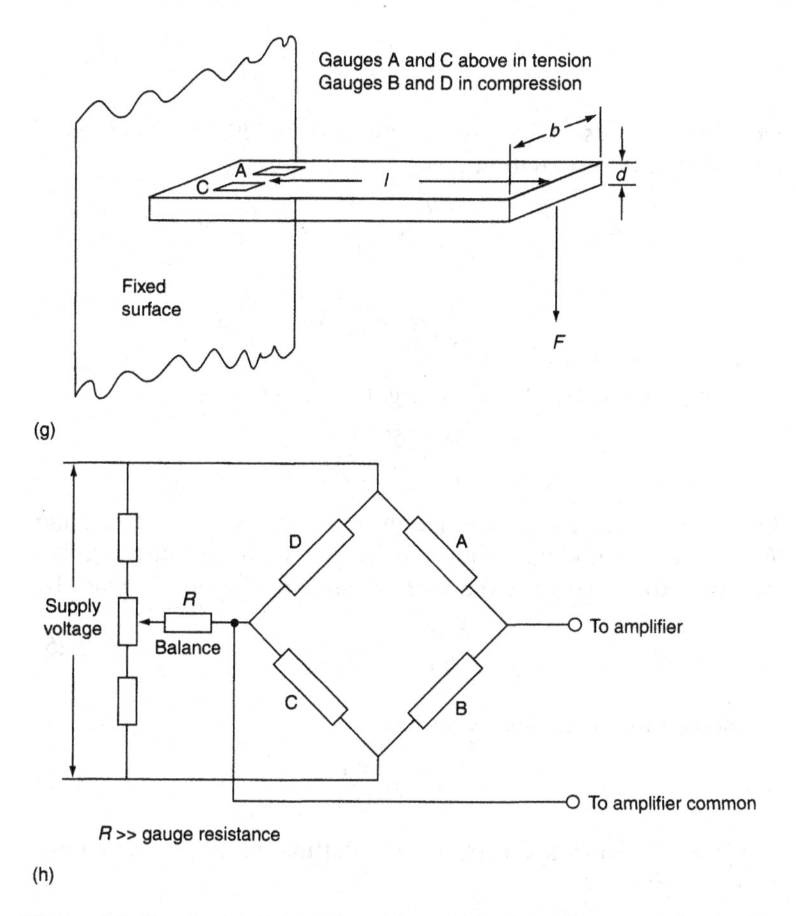

Figure 3.5 *Continued. (g) Application of (a) to bending a cantilever; (h) circuit.*

3.6.3 Gauge factor

The fractional change in resistance of a strain gauge is greater than the strain, because the wire or foil strip not only becomes longer, but also thinner and its resistivity increases. The **gauge factor** G is defined as

$$\frac{\delta R/R}{\delta l/l} \tag{3.24}$$

It is derived as follows:

$$R = \frac{\rho l}{wt}$$

where ρ is the resistivity, l the length, w the width and t the thickness. Partial differentiation gives

$$\delta R = \frac{\partial R}{\partial \rho}\delta\rho + \frac{\partial R}{\partial l}\delta l + \frac{\partial R}{\partial w}\delta w + \frac{\partial R}{\partial t}\mathrm{d}t$$

$$= \frac{l}{wt}\delta\rho + \frac{\rho}{wt}\delta l - \frac{\rho l}{w^2 t}\delta w - \frac{\rho l}{wt^2}\mathrm{d}t$$

Dividing the left by R and the right by $\rho l/wt$ gives

$$\frac{\delta R}{R} = \frac{\delta\rho}{\rho} + \frac{\delta l}{l} - \frac{\delta w}{w} - \frac{\mathrm{d}t}{t}$$

Let α_L be the fractional change in resistivity per unit stress, and E be Young's modulus. Then the fractional change in resistivity per unit strain, which is the **piezoresistive coefficient**, is given by

$$\frac{\delta\rho/\rho}{\delta l/l} = E\alpha_L \tag{3.25}$$

Poisson's ratio, ν, is defined as

$$\frac{\delta w/w}{\delta l/l} = \frac{\delta t/t}{\delta l/l} \tag{3.26}$$

For metals, Poisson's ratio ≈ 0.33. Putting all these equations together gives

$$G = 1 + E\alpha_L + 2\nu \tag{3.27}$$

The gauge factor is approximately two for foil strain gauges but can be in the range 50–200 for silicon gauges, because they have a much larger value of α_L. This makes silicon gauges much more sensitive to strain than foil gauges, but also correspondingly more sensitive to temperature.

3.6.4 Circuit layout and temperature compensation

Temperature rises affect strain gauges in three ways:

- the resistivity of the gauge foil increases;
- the gauge foil expands;
- the material to which it is attached also expands.

To a limited extent the last two compensate each other. The change in resistance due to strain is small anyway, less than 1%, so gauges are always used in pairs or fours in a bridge circuit (Figure 3.5(h)). If A is an active gauge in tension, B will be either in compression or unstressed and providing only temperature compensation. Gauges C and D could be resistors, but for maximum sensitivity C will be in tension like A, and D will be like B. As well as compensating for temperature, this circuit can separate the effects of tension and bending as shown in Figures 3.5(d) and (g). As the resistances of the four gauges may not be exactly equal, a balance adjustment is also needed. The resistance R is much greater than the gauge resistance. The sensitivity of a strain gauge sensor is directly proportional to the energizing current, but if the current is too large the heating it causes will upset the balance. The gauge manufacturers recommend a maximum current for each type of gauge.

3.6.5 Signal conditioning

A strain gauge bridge requires a voltage source which does not drift and is not noisy. Even with the maximum allowable current measuring a large strain, the output of a strain gauge bridge is only a few millivolts and it may be much less, so an instrumentation amplifier is required to bring the output up to an appropriate level for data acquisition. These are described on page 66. Plug-in data acquisition boards for desk-top computers can be obtained, including amplifiers and circuits for bridge exitation. The voltage from the computer power supply will probably be too noisy to be used directly.

3.7 Linear variable differential transformer

The linear variable differential transformer, invariably known as the LVDT, is a very useful device for measuring displacements from a range of a few millimetres and a resolution of micrometres to a range of tens of centimetres.

The construction of the device is illustrated by the cross-section shown in Figure 3.6(a). Three coils are wound on a nonmagnetic former with a magnetic outer screen. The central coil is energized and is the primary of the transformer. LVDTs are designed to be energized at a particular frequency in the range from 50 Hz to 20 kHz. The two end coils are secondaries and a voltage is

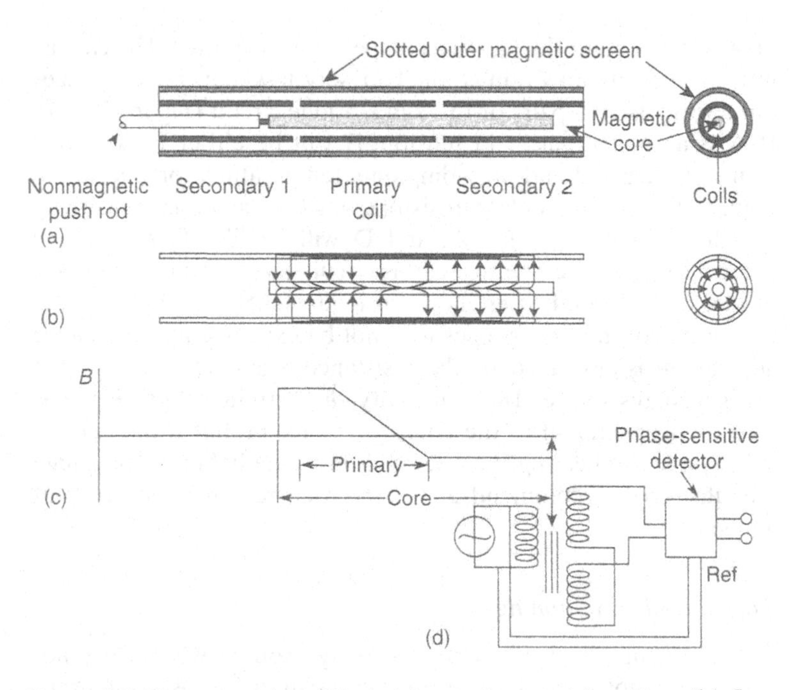

Figure 3.6 *Linear variable differential transformer: (a) section; (b) flux path; (c) flux density in gap; (d) circuit.*

induced in each of them. Down the centre of the coils is the core or armature, which is made of magnetic material. The core may be mounted in bearings or it may be quite free, but it can move in the axial direction. It is attached to the object whose movement is to be measured by a nonmagnetic push rod. The flux generated by the current in the primary coil passes along the core, across the gap to the outer screen and back to the primary, Figure 3.6(b). In crossing the gap from the core to the outer shield the flux links the secondary coils and induces an electromagnetic field (emf) in each. The emf in each secondary is proportional to the length of the core inside it. The secondary coils are connected back to back so that the a.c. output is zero when the core is central and maximum when it is at either end. In order to tell which way the core is displaced, the output is connected to a phase-sensitive detector with the energizing voltage as reference. This circuit is fully described on page 82. The velocity of movements which can be measured is limited by the carrier frequency. As a rule of thumb, the highest frequency of oscillatory movement which can be measured is one-tenth of the carrier frequency. Although the output signal varies according to the design of a particular LVDT and the excitation voltage, in general it is an order greater than the output of a strain gauge bridge.

3.8 Piezoelectric sensors

3.8.1 Piezoelectricity

The name piezo comes from the Greek word for 'press'. Certain crystals develop charges on their surfaces when they are pressed. These include Rochelle salt (sodium potassium tartrate), tourmaline and quartz. Quartz is silicon oxide and crystallizes in hexagonal rods. To make sensors, slices are cut from the crystals. Depending on how the plane of the slice relates to the axes of the crystal, pressure applied across two faces may produce a charge on faces normal to the pressure (X-cut) or on the faces parallel to the pressure (Y-cut) as shown in Figure 3.7(a). A very simplified explanation of how deforming the crystal produces surface charges in quartz is shown in Figures 3.7(b) and (c). Natural quartz is widely distributed, but for scientific applications crystals are grown synthetically. Synthetic ceramics have been developed with piezoelectric properties. These are ferroelectric materials which are not intrinsically piezoelectric, but become so by the application of a very strong electrostatic field during manufacture. Many materials have been developed for particular purposes. Some have a wide working temperature range, others have a very high sensitivity, but in general the sensitivity of synthetic materials is an order greater than quartz.

For a detailed description of the theory see Jaffe and Berlincourt (1965), and for more practical accounts see Pennington (1965), Doebelin (1990) and Holman (1993).

A crystal without defects does not absorb energy, so it is a perfect electromechanical transducer in either direction. Piezoelectric sensors are self-generating and do not require a power supply, although power is required for the auxiliary electronics.

Piezoelectric constants

Table 3.2 summarizes the symbols used for easy reference. The left-hand side shows the material constants per unit area or per

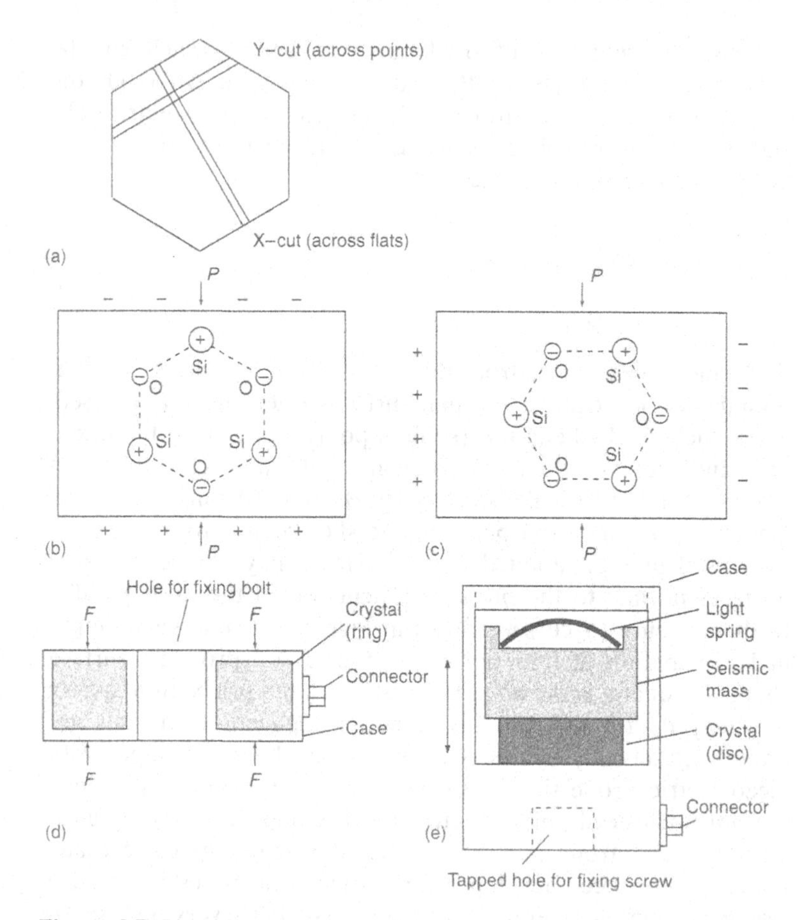

Figure 3.7 *Piezoelectric sensors: (a) slices from a hexagonal crystal; (b) simplified explanation of piezoelectricity in quartz, X-cut; (c) as (b) but Y-cut; (d) load washer (section); (e) accelerometer (section).*

unit thickness as the case may be, and the right-hand side shows the parameters as they are specified for a sensor of area A and thickness t. The electrical parameters apply when considering the material as a dielectric; the mechanical parameters apply to it as an elastic solid which changes its thickness by x; and the piezoelectric parameters relate the piezoelectric output to both electrical and mechanical inputs.

The performance of a piezoelectric material or a sensor is described by a number of constants. Two of particular importance

Table 3.2 *Parameters of piezoelectric materials and sensors*

Material		*Sensor*	
Parameter	*Symbol (unit)*	*Parameter*	*Symbol (unit)*
Electrical parameters			
Electric field	E (V m^{-1})	Voltage	V (volts)
Electric displacement	D (C m^{-2})	Charge	Q (coulombs)
Permittivity of space	ϵ_0 (F m^{-1})	Universal constant	ϵ_0 (F m^{-1})
Relative permittivity	ϵ_r (none)	Capacitance	C (farads)
Energy per unit cube	ED (J m^{-3})	Energy	QV (joules)
Mechanical parameters			
Stress	P (N m^{-2})	Force	F (newtons, N)
Strain	S (none)	Displacement	x (metres)
Young's modulus	Y (N m^{-2} = Pa)	Stiffness	F/x (N m^{-1})
Energy per unit cube	PS (J m^{-3})	Energy	Fx (joules)
Piezoelectric parameters			
Charge sensitivity	$d = D/P$ C N^{-1} = m V^{-1} = Q/F	(coulombs newton^{-1})	
Voltage sensitivity	$g = E/P$ V m N^{-1} = m^2 C^{-1} = $VA/(Ft)$	(voltmeters newton^{-1})	

are the charge constant and the voltage constant. The charge constants d for a unit cube and for a sensor are, respectively,

$$d = \frac{D}{P} \quad \text{and} \quad d = \frac{Q}{F} \tag{3.28}$$

The units of d are C N^{-1} = m V^{-1}. If it seems odd that these units should be the same, note that CN^{-1}/mV^{-1} = CV/Nm and that both coulomb.volts and newton.metres = joules, and measure energy.

The voltage constants g for a unit cube and for a sensor are

$$g = \frac{E}{P} \quad \text{and} \quad g = \frac{V}{F} \tag{3.29}$$

respectively. The units of g are V m N^{-1} = m^2 C^{-1}. The constants g and d are written with two subscripts, i and j, where i indicates the electrical direction and j the mechanical direction. Their values are 1 for the x-axis, 2 for the y-axis and 3 for the z-axis. In ferroelectric materials, the z-axis is the direction of polarization. The piezoelectric constants are related by the equation

$$g = \frac{d}{\varepsilon_0 \, \varepsilon_r}$$

The mechanical parameters are related by Young's modulus:

$$Y = \frac{P}{S} \tag{3.30}$$

This unfortunately has the standard symbol E, so Y has been used here to avoid confusion with the electric field, E.

The electrical parameters are related by permittivity:

$$\frac{D}{E} = \varepsilon_0 \, \varepsilon_r \tag{3.31}$$

where ε_0 is the permittivity of free space = 8.854×10^{-12} F m^{-1}, and ε_r is the relative permittivity of the material.

The piezoelectric equation can take many forms depending on whether the crystal is connected to a high impedance or a low one and whether it is constrained to produce a force or free to expand or contract. Also, the equations exist in corresponding forms for a unit cube of the material and for a complete sensor.

As an example, consider pressure applied to a unit cube of piezoelectric material. The crystal behaves both as a spring and as a capacitor.

The piezoelectric effect (3.28) gives

$$D = dP$$

due to the stress. Also, from (3.31),

$$D = \varepsilon_0 \, \varepsilon_r \, E$$

due to the electric field. The combined effect is

$$D = \varepsilon_0 \, \varepsilon_r{}^{P} E + dP \tag{3.32}$$

The superscripts indicate the conditions, i.e. $\varepsilon_0 \varepsilon_r{}^{P}$ is the permittivity at constant stress. Algebraic rearrangement of (3.32) gives

$$E = \frac{D}{\varepsilon_0 \, \varepsilon_r{}^{P}} - gP \tag{3.33}$$

For a sensor based on a crystal of area A and thickness t, the corresponding equations are

$$Q = CV + dF \tag{3.34}$$

and

$$V = \frac{Q}{C} - \frac{gFt}{A} \tag{3.35}$$

There are many more permutations of these equations, and which version is appropriate depends on the external conditions on the output side. A force sensor working into an instrumentation amplifier with a high input impedance can develop voltage, but no charge flows out of the sensor. If it is working into a charge amplifier, all the charge flows out and no voltage is left across the sensor. (Voltage and charge amplifiers are explained on pages 66 and 70.)

As mentioned above, piezoelectric materials can transduce electrical energy to mechanical as well as vice versa. As a mechanical-to-electrical transducer, the strain S can be found by a similar argument:

$$S = \mathrm{d}E + \frac{P}{Y^{E}} \tag{3.36}$$

$$S = gD + \frac{P}{Y^{D}} \tag{3.37}$$

3.8.2 Practical piezoelectric sensors

Strain gauges are used to measure force via the stiffness of the material on which they are mounted, but piezoelectric sensors are intrinsically sensitive to compressive force. They are made into **load washers** with which to measure compressive force (Figure 3.7(d)), prestressed to make **force links** with which to measure force in tension or compression, combined with a seismic mass to make accelerometers (Figure 3.7(e)) or exposed to fluid to make pressure sensors. A microphone is a specialized pressure sensor. They are widely used in mechanical engineering to measure shock and vibration. The principal advantages of piezo sensors include the following:

- The resolution is a very small fraction of the maximum measurand, although the linearity is only of the order of 1%. For example, one type of quartz sensor will measure up to 20 000 N with a resolution of 0.02 N.
- The high-frequency response of a piezoelectric sensor is typically flat up to about 10 kHz, and somewhere above that there is a sharp peak at the natural frequency of the crystal and its mounting.

The main disadvantage of piezoelectric sensors is that they have no static response, because charge inevitably leaks away. With a specially designed high-impedance input amplifier, guard electrodes and very clean surfaces, it is possible to obtain a time constant of several minutes from a quartz sensor. Ceramic sensors have shorter time constants. If a static response is essential, a piezoresistive or strain-gauge type of sensor is preferable.

To avoid errors due to cable and amplifier input capacitance, piezoelectric sensors are used with **charge amplifiers**, which are described on page 70.

3.8.3 Ultrasonics

As mentioned above, the piezoelectric effect is reversible, and crystals are also used as ultrasonic transducers to detect cracks, measure the thickness of materials accessible from one side only and to measure depth of water, i.e. SONAR.

3.8.4 Frequency control

Frequency control is probably the commonest application of piezoelectricity. The crystal is in mechanical resonance but appears at the electrical terminals as a very sharply tuned circuit. Crystals have a much higher Q (about 10^6) than can be obtained by purely electrical means and slices for this application are cut from the crystal in the direction which gives the lowest temperature coefficient.

3.9 Temperature sensors

One of the most commonly measured parameters in industry and elsewhere is temperature, and a large number of physical parameters are temperature dependent.

The theory of temperature scales is outlined in Chapter 2. A scale is necessary because not all the temperature-dependent parameters vary in quite the same way. The one which comes closest to the theoretical definition of temperature over most of the range is the resistivity of pure platinum. Quinn (1990) has written a comprehensive and authoritative survey of both the theory and practice of temperature measurement.

3.9.1 Platinum resistance thermometers

Many types of platinum resistance thermometer are manufactured, ranging from those used in standards laboratories to mass-produced wire and foil sensors for industry (Atherton and Fitton, 1989). The abbreviation PRT is used in the UK and RTD (resistance temperature detector) in the USA. The highest accuracy PRTs comprise a lightly supported wire coil mounted in a quartz protective tube. The support must be gentle or else the wire coil behaves partially as a strain gauge. The platinum must be pure because contamination reduces the temperature coefficient, and minimum values at certain temperatures are specified if a thermometer is to be used for calibrating other thermometers.

In industrial thermometers, the platinum wire is enclosed in a ceramic tube and protected by a stainless steel sheath, and possibly also a steel well or pocket. Another type consists of a grid of platinum foil on an alumina substrate (alumina is one of the few materials which is an electrical insulator but a thermal conductor).

The relationship between temperature and resistance is adequately described by the Callendar equation below. In order to solve this equation it is usually treated as a linear function with a small correction for nonlinearity. The thermometer is calibrated at 0 °C and 100 °C, and the mean temperature coefficient for this interval is designated α:

$$\alpha = \frac{R_{100} - R_0}{100R_0} \tag{3.38}$$

Another calibration at the freezing point of zinc gives the linearity correction δ:

$$R_T = R_0\left\{1 + \alpha\left[T - \delta\left(\frac{T}{100}\right)\left(\frac{T}{100} - 1\right)\right]\right\} \tag{3.39}$$

where T is the temperature (°C) and R_T is the resistance at T.

For temperatures below 0 °C a further correction β is made, and this becomes the Callendar–Van Dusen equation:

$$R_T = R_0\left\{1 + \alpha\left[T - \delta\left(\frac{T}{100}\right)\left(\frac{T}{100} - 1\right) - \beta\left(\frac{T}{100}\right)^3\left(\frac{T}{100} - 1\right)\right]\right\} \tag{3.40}$$

Most industrial sensors are not individually calibrated, but are required to meet a standard specification, such as Pt100 which is defined in BS 1904:1984, DIN 43760 and IEC 751:1983. All these require that R_0, the resistance at 0 °C, is 100 Ω and that $\alpha = 0.003850$ with a tolerance of $\pm 0.15 + 0.002 \times T$°C for class A and $\pm 0.3 + 0.005 \times T$°C for class B. Tables of the resistance for each temperature which take the nonlinearity into account are published in the standards documents and elsewhere (Quinn, 1990). An alternative standard used in the USA and Japan is D100. This specifies $\alpha = 0.003916$. The accuracy of these thermometers is not sufficient to justify further correction to ITS-90.

3.9.2 Signal conditioning for platinum resistance thermometers

As with strain gauges, it is necessary to pass a current through the sensor to detect the change in its resistance, but the current must be sufficiently small for self-heating to be negligible. The normal value is 1 mA in a 100 Ω sensor. This gives a sensitivity of $0.00385 \times 100 \times 0.001 = 0.000\,385$ V °C^{-1}, about 0.4 mV °C^{-1}.

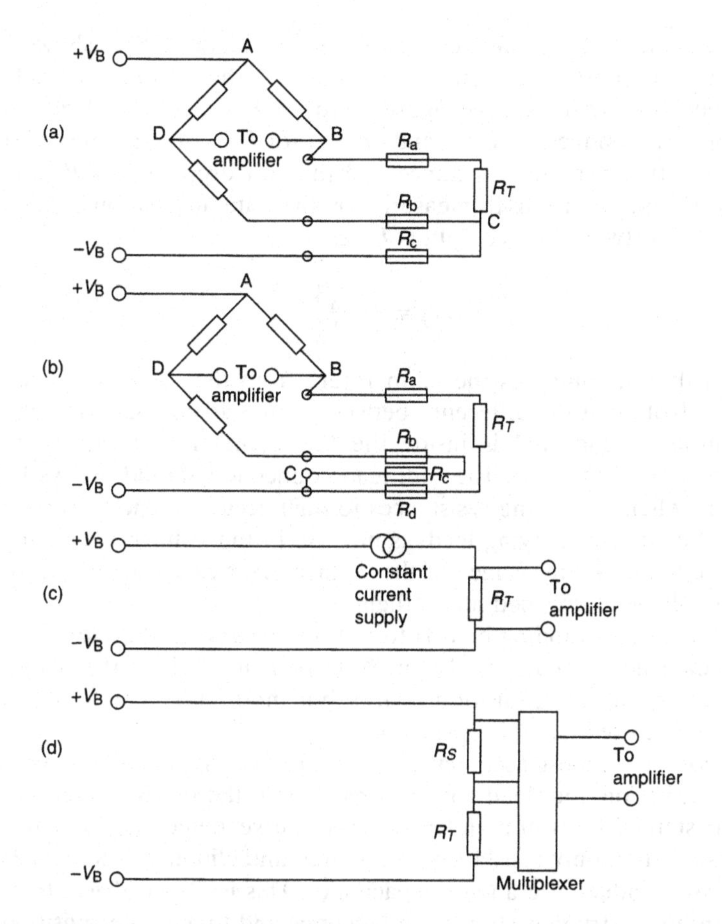

Figure 3.8 *Circuits for platinum resistance thermometers: (a) three-lead compensated bridge; (b) four-lead compensated bridge; (c) four-lead connected, constant current; (d) four-lead potentiometric.*

It is important to avoid errors due to thermoelectric emfs and to changes in the resistance of the leads. Thermoelectric emfs are cancelled by measuring the sensor resistance twice with current flowing in opposite directions or using low frequency a.c.

The possible circuit arrangements are shown in Figure 3.8. A three-lead bridge circuit is shown in Figure 3.8(a). The resistance R_a of one lead appears between B and C, and R_b appears between C and D. Figure 3.8(b) shows a four-lead compensated bridge, and the resistances of two leads, R_a and R_d, appear between B and C, and R_b and R_c appear between C and D. Figure 3.8(c) shows a four-lead constant current circuit, which is the simplest for data acquisition systems; and Figure 3.8(d) shows a four-lead potentiometric connection. The standard resistor R_S is of the same order as the thermometer resistance and the volt drops across R_S and R_T; V_S and T_T are both measured by the data acquisition system.

The software first computes R_T as

$$R_T = \int \frac{R_S V_T}{V_S}$$

and then it computes the temperature from the Callendar equation. Note that the difference between four-lead compensated and four-lead connected is inside the thermometer and cannot be changed by the user. The four-lead connection should always be used when measuring resistances to high accuracy. The resistance of the current-carrying leads is not within the voltage measuring circuit, and the resistance of the voltage leads carry no current so no voltage is dropped along them.

Signal-conditioning boards to connect resistance thermometers to data acquisition boards for PCs are marketed by the manufacturers of the acquisition boards, but these may be external to the PC to reduce noise problems.

For the highest accuracy, a special bridge circuit is used with a.c. exitation, four-lead connections to both the thermometer and the standard resistor and a phase-sensitive detector. The latter ensures that only resistance is measured and eliminates the effects of stray inductance and/or capacitance. This is a specialized, free-standing instrument, but it can be connected to a data acquisition system via the general purpose interface (GPIB).

3.9.3 Thermistors

Thermistors are temperature-sensitive oxides of manganese and nickel. They come in all sorts of useful shapes and sizes, and both positive and negative temperature coefficients, although negative is more common.

The resistance of a thermistor is related to the temperature by the equation

$$R_T = R_0 \exp[-B(T^{-1} - T_0^{-1})] \tag{3.41}$$

where R_T is the resistance at temperature T and R_0 the resistance at temperature T_0.

The value of B depends on the composition of the material, and is of the order of 4000. The temperature coefficient α is given by

$$\alpha = \frac{1}{R}\frac{\mathrm{d}R}{\mathrm{d}T} = \frac{-B}{T^2} \tag{3.42}$$

So, although thermistors are much more sensitive than PRTs, they are much less linear and also less stable.

3.9.4 Thermocouples

Whenever there is a temperature gradient along a conductor, there is also a potential gradient. This is the **Seebeck effect**. The magnitude of the potential gradient depends on the material of the conductor. Thus if a circuit made of a uniform material is not at a uniform temperature, the emf between the cold part and the hot part on one side is exactly matched by the emf on the other side of the circuit from hot to cold. However, if the circuit is made of two materials A and B, and one junction is hot and the other is cold, then the emf in A from cold to hot is not matched by the emf in B. The result is a net emf which may be used to measure the temperature difference. This is a **thermocouple**. Conversely, when current flows from one material to another, heat is absorbed or released at the junction according to the direction of current flow. This is the **Peltier effect**. In metals, the Peltier cooling is always smaller than the Joule heating (i^2R), but in semiconductors the Peltier coefficient may be so large that the junction is cooled, and this has practical applications. Heat is also absorbed or released by a current flowing through a conductor made of uniform material if there is a temperature gradient along the conductor. This is the **Thomson effect** or **Kelvin effect**, and is in addition to the Joule heating.

There are a number of material pairs which are commonly used for practical thermocouples and are identified by code letters. Table 3.3 summarizes their characteristics. It can be seen that the thermoelectric or Seebeck emf is not strictly proportional to the temperature difference and detailed tables are published relating emf to the temperature difference for all these materials (Quinn, 1990; BS 4937 (1981) and IEC 584-1; Kaye, 1986).

Thermocouples are available in various shapes and sizes for different purposes. Bare wire couples have the fastest response, but are fragile. For measuring the temperature of a liquid, the

Table 3.3 *Common thermocouple materials*

Letter code	*Materials*	*Usable range (°C)*	*emf, cold junction 0 °C (mV)*		
			20 °C	*100 °C*	*400 °C*
T	Copper–Constantan Cu–(60% Cu, 40% Ni)	–200 to 400	0.79	4.28	20.87
J	Iron–Constantan Fe–(60% Cu, 40% Ni)	–200 to 1200	1.02	5.27	21.85
K	Chromel–Alumel (90% Ni, 10% Cr)–(94% Ni, 2% Al etc.)	–300 to 1300	0.80	4.10	16.40
S	Platinum–Platinum 10% Rhodium	–100 to 1700	0.11	0.65	3.26
R	Platinum–Platinum 13% Rhodium	–100 to 1750	0.11	0.65	3.41

thermocouple may be encased in a steel sheath, as for a resistance thermometer, and its response time is then limited as described above. **Tab** thermocouples consist of two strips of appropriate metals welded together side by side to form a tab which can be pressed against a solid surface to measure its temperature.

3.9.5 Signal conditioning for thermocouples

It can be seen from Table 3.3 that the output of a thermocouple is in the millivolt range and therefore needs amplification before the A/D converter. Unlike a PRT, a thermocouple also needs linearization and cold junction compensation. Linearization may be done in hardware in the input circuit, but more satisfactorily in software. As each input board in a PC-based system needs its own driver program, this is where the linearization is done, either by an equation or a look-up table. There is often provision in the high-level program for the user to tell the driver which type of thermocouple is connected, and the driver selects the appropriate linearization routine.

The cold junction of the thermocouple is at the terminal block where it is connected to the signal-conditioning board, and the thermocouple emf measures only the difference between the block and the hot junction. Unless the hot junction temperature is so large that changes in the temperature of the terminal block may be neglected, it must be measured with a PRT, thermistor or integrated circuit sensor, so that the true hot junction temperature can be computed. The AC1226 is a specialized IC for this purpose. If more than one thermocouple is in use, it is recommended that they are all terminated in an isothermal block whose temperature is the reference and is measured.

Signal-conditioning boards to connect thermocouples to data acquisition boards for PCs are marketed by the manufacturers of the acquisition boards, but they may be external to the PC to reduce noise problems.

3.9.6 Integrated circuit temperature sensors

The base–emitter voltage of a transistor varies with temperature. This effect is utilized in devices such as the AD590 or AD592 from Analog Devices Inc., to make temperature sensors with a fixed sensitivity of 1 $\mu V °C^{-1}$. These devices may be obtained in the normal IC packages or in a steel sheath. They are very convenient to use, but the range is the normal industrial (–40°C to +85°C) or military (–55°C to +125°C) range for ICs, which is more limited than PRTs or thermistors. The AD590 has a maximum nonlinearity of 1.5°C and the AD592 works over the more limited industrial range and has a maximum nonlinearity of 0.35°C.

The sensitivity of the base–emitter voltage to temperature is utilized for temperature compensation in a bandgap voltage reference, as seen in Chapter 7.

3.9.7 Radiation sensors

The temperature of very hot or moving bodies can be measured only by measuring the heat radiated from them. The relationship between radiated power, its wavelength and the absolute temperature of the hot body, is summarized by three laws:

1. *Stefan–Boltzmann law* The total radiation (all wavelengths) is proportional to T^4, where T is the absolute temperature, i.e. the temperature in kelvin. If a body at temperature T_1 is in an enclosure at temperature T_2, the net radiation from the hotter to the cooler body is:

$$W \propto (T_1^4 - T_2^4) \tag{3.43}$$

2. *Planck's equation* This relates the power, in watts, radiated from 1 m^2 of a 'black body' between the wavelengths of λ and $\lambda + d\lambda$, Quinn (1990):

$$M_\lambda d\lambda = 2\pi hc^2\lambda^{-5} [\exp(hc/k\lambda T) - 1]^{-1} d\lambda \qquad (3.44)$$

where M_λ is the spectral exitance at wavelength λ, h is Planck's constant (6.636×10^{-34} J s), c is the velocity of light (2.997×10^{-8} m s^{-1}), and k is Boltzmann's constant (1.381×10^{-23} J K^{-1}).

The spectrum of the radiation from a black body at various temperatures follows equation (3.44) and is shown in Figure 3.9. A black body is one which absorbs all radiation falling on it without reflecting any, and conversely radiates the maximum

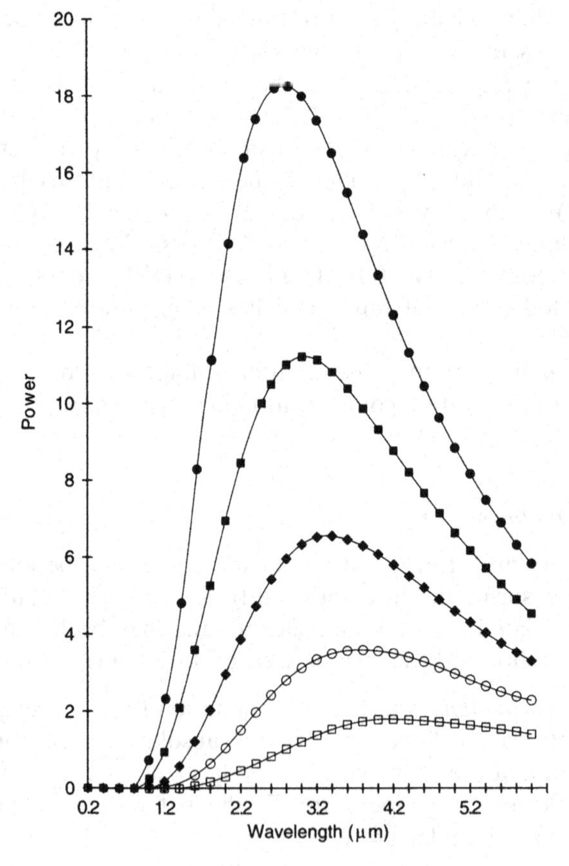

Figure 3.9 *Radiation from a black body: Planck's law: (●) 800 °C; (■) 700 °C; (◆) 600 °C; (○) 500 °C; (□) 400 °C.*

power for its temperature. The terminology is slightly confusing as a black body at a sufficiently high temperature will glow red or even white.

3. *Wien's displacement law* A line through the maxima of the Planck equation curves follows this law, which relates to the wavelength of the maximum radiation, λ_{max} to the temperature T, in kelvin:

$$\lambda_{max} \propto T \tag{3.45}$$

Silicon photodiodes are sensitive to radiation from 0.39 to 1.15 μm with maximum sensitivity at approximately 0.8 μm. These are very useful for detecting visible radiation but, for temperature sensing, photoresistors which are sensitive to infrared are more useful. The combination of the spectrum of emitted radiation and limited spectral sensitivity of radiation detectors means that they are extremely nonlinear, the sensitivity increasing rapidly with temperature. There are further complications in that the radiated power depends not only on the temperature, but also on the condition of the surface, quantified by the emissivity. Nevertheless in engineering applications where the hot body is moving, and in the measurement of furnace temperatures, radiation sensors are the only ones available.

3.10 Future developments

Most of the sensors described above have been available for many years, but semiconductor sensors are becoming increasingly common. In general they are more sensitive than the older types. Temperature compensation as described above for strain gauges is particularly important. At one time all strain gauges and resistance thermometers were manufactured from thin wire, but now most strain gauges and some thermometers are made from foil which is screen printed onto an insulating substrate.

3.10.1 Microsensors

The older sensor designs are labour intensive to manufacture, whereas semiconductors lend themselves to mass production and miniaturization.

The use of silicon for transistors and integrated circuits has resulted in sophisticated techniques and machinery becoming available for mass producing very pure silicon and doping it with controlled amounts of impurities, and also for machining it on a microscopic scale. These techniques are now applied to the design and manufacture of very tiny sensors, some of which can be mass produced very cheaply. For example, an accelerometer can be made entirely out of silicon. The seismic mass is a block supported within the frame by a thin strip which forms a springy cantilever. The four piezoresistive strain gauges are formed within the silicon of the cantilever by ion implantation. The dimensions of the seismic mass for a conventional accelerometer are quoted in millimetres and grams, but for a microaccelerometer, micrometre and microgram are more appropriate. Another example is a pressure sensor in which the diaphragm is a thin slice of silicon and the strain gauges are formed within it in the appropriate places. For more detailed descriptions see Göpel *et al.* (1989), Middlehoek and Audet (1989), Gardner (1994 and 1995), or Rudolf *et al.* (1995).

3.10.2 Smart sensors

The next stage of sensor development is **smart sensors**, defined by the IEEE as 'a device with built-in intelligence, whether apparent to the user or not'. As the digitization and the microprocessor are included within the sensor package, these are beyond the scope of this book, but further information is given by Brignell and White (1994), Wolfenbuttel (1995) and Frank (1996).

3.11 Examples

Strain gauges are frequently used to measure the force in part of a machine or other structure. The first two examples show how to calculate the sensitivity of a gauge used in this way. Example 3.1 applies to a tube in torsion, i.e. twisting, and Example 3.2 applies to a bending force on a cantilever. The relationships between force (or torque) and strain for other shapes will be found in any standard mechanical engineering text.

Example 3.1

A strain gauge is mounted on the outer surface of a metal tube at an angle α to the longitudinal direction of the tube. From first principles calculate the strain in the gauge when a torsional moment T is applied to the tube.

A steel tube 10 mm outside diameter and 1 mm wall thickness is fitted with four gauges configured to measure torsion. The bridge voltage is 6 V and the gauge factor is 2.10. The measured output is 12.5 mV. Calculate the torque applied to the tube, given that the shear modulus of steel G is 82.1×10^9 N m^{-1}.

Solution

Figure 3.10(a) represents one gauge applied to the outside of a cylinder at an angle α to the axis. The bridge is completed by three more gauges as shown in Figures 3.10(c) and (d).

When the torque T is applied to the cylinder, Q rotates relative to P so that y becomes $y + \delta y$. Then, the shear strain at the surface, S_s is given by

$$S_s = \frac{\delta y}{x} = \frac{\delta y}{l_G \cos \alpha}$$

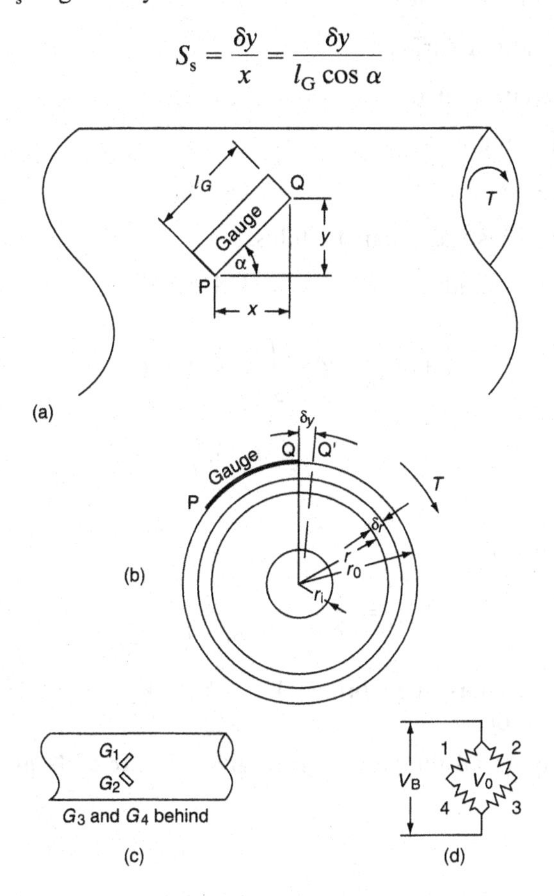

Figure 3.10 *Strain gauges to measure torque.*

increase in the length of the gauge $\delta l_G = \delta y \sin \alpha$

strain in the gauge

$$S_G = \frac{\delta y \sin \alpha}{l_G} = S_s \sin \alpha \cos \alpha$$

when $\alpha = 45°$, $S_G = ½S_S$.

To find S_s, we consider an annulus of width δr and radius r, calculate the shear force and hence the torque on the annulus. Then we integrate from the inner to the outer radius to find the total torque. Figure 3.10(b) shows:

shear strain at centre $= 0$

shear strain at surface $= S_s$

shear strain S_r at radius r $= S_s(r/r_0)$

area of annulus A_r $= 2\pi r\ \delta r$

shear force on annulus $= GS_rA_r$,

where G is the shear modulus,

torque on annulus $= GS_rA_rr$

$$\text{total torque } T = \int_{r_i}^{r_0} GS_rA_rr\, \mathrm{d}r$$

$$= \frac{\pi GS_s}{2r_0}(r_0^4 - r_i^4) \tag{3.46}$$

strain in gauge

$$S_G = \frac{S_s}{2} = \frac{Tr_0}{\pi G(r_0^4 - r_i^4)}$$

output of four-gauge bridge $V_0 = S_G G_F V_B$, where G_F is the gauge factor.

Substituting all the values given gives $T = 18.88$ N m.

Example 3.2

A cantilever with strain gauges mounted on the upper and lower surfaces is to be used to measure weight in an industrial environment. The width of the cantilever is 40 mm and a mass of 1 kg is placed 200 mm from the gauges. There may be a temperature difference of 1°C between the upper and lower surfaces causing a spurious signal. Calculate the maximum thickness of the cantilever so that the strain produced by the 1 kg mass is at least 25 times the spurious signal, given the following parameters:

coefficient of linear expansion of cantilever material	$= 15 \times 10^{-6}$
coefficient of linear expansion of strain gauge foil	$= 17 \times 10^{-6}$
temperature coefficient of resistivity of gauge foil	$= 20 \times 10^{-6}$
gauge factor	$= 2.1$
Young's modulus for cantilever material E	$= 210 \times 10^{9}$

The strain S_P at a point P on the surface of a cantilever is given by

$$S_P = \frac{6l}{Ebd^2} F \tag{3.47}$$

where E is Young's modulus, b is the breadth of the cantilever, d is the depth of cantilever, and F the force parallel to d at distance l from P (Figure 3.5(g)).

Solution

A mass of 1 kg applies a force of 9.81 N to the cantilever. The strain produced in the upper gauge at P, 200 mm away, is

$$S_P = \frac{6 \times 0.2}{210 \times 10^9 \times 0.04 \times d^2} \times 9.81 = \frac{1.401 \times 10^{-9}}{d^2}$$

The same strain, but in the opposite direction, is produced in the lower gauge, to give a differential strain of $2 \times S_P$.

The differential change in resistance due to strain is

$$\delta R_P = \frac{2 \times 2.1 \times 1.401 \times 10^{-9}}{d^2}$$

$$= \frac{5.886 \times 10^{-9}}{d^2}$$

The effect of the temperature difference is now calculated. The values given for the expansion of the cantilever and the expansion of the foil are both expressed as a fraction of the original length, i.e. strain. The foil expands more than the metal to which it is attached, equivalent to a compressive strain of 2×10^{-6} per °C, which reduces the resistance.

The differential decrease in resistance due to expansion of metal and foil is $2.1 \times 2 \times 10^{-6}$, and the differential increase in resistance due to increase in resistivity is 20×10^{-6} per °C.

The combined effect of the 1°C temperature difference is the change of resistance:

$$\delta R_T = (20 - 2.1 \times 2) \times 10^{-6} = 15.8 \times 10^{-6}$$

The criterion in the question is that $\delta R_P \geqslant 25 \times \delta R_T$, i.e.

$$\frac{5.886 \times 10^{-9}}{d^2} \geqslant 25 \times 15.8 \times 10^{-6}$$

Hence,

$$d^2 \leqslant 7.449 \times 10^{-6}$$

and the maximum thickness of cantilever $d = 3.86$ mm.

Example 3.3

The Callendar equation (3.39) makes it simple to calculate the resistance for a known temperature, but more usually we measure the resistance and wish to find the temperature. Tables are published for all the popular types of thermometer. This example shows how to use the Callendar equation iteratively in case suitable tables are not available.

A platinum resistance thermometer is supplied with the following calibration data in standard notation:

$$R_0 = 100.000\ \Omega$$

$$R_{100} = 138.394\ \Omega$$

$$\delta = 1.497\ ^\circ\mathrm{C}$$

Using Callendar's equation calculate the temperature to the nearest 0.01°C when $R = 150.000\ \Omega$.

Solution

Callendar's equation (3.39) is

$$R_T = R_0\{1 + \alpha[T - \delta(0.01T)(0.01T - 1)]\}$$

where T is the temperature in °C, and R_T is the resistance at T.

The first estimate of temperature T' with no correction for nonlinearity is

$$T' = \frac{R_T - R_0}{R_{100} - R_0} \times 100 = \frac{50.000}{38.394} \times 100 = 130.229$$

and the nonlinearity correction is

$$T - T' = \delta(0.01T)(0.01T - 1)$$

The second estimate with nonlinearity correction is

$$\begin{aligned} T &= 130.229 + 1.497 \times 1.30229 \times 0.30229 \\ &= 130.229 + 0.589 \\ &= 130.818 \end{aligned}$$

Third estimate:

$$\begin{aligned} T &= 130.229 + 1.497 \times 1.30818 \times 0.30818 \\ &= 130.229 + 0.604 \\ &= 130.833 \end{aligned}$$

Fourth estimate:

$$\begin{aligned} T &= 130.229 + 1.497 \times 1.30833 \times 0.30833 \\ &= 130.229 + 0.604 \\ &= 130.833 \end{aligned}$$

As this is the same as the previous estimate we can stop and state that, to the nearest 0.01 °C, the temperature = 130.83 °C.

CHAPTER 4

Signal conditioning

4.1 General

The output of a sensor is an analogue of the quantity being measured, the measurand. Depending on the type of sensor, the output may be a voltage (mV or μV), a current or possibly another parameter, and it may have the same frequency as the measurand or it may have a carrier frequency modulated by the measurand. Most A/D converters require an input of a few volts at the same frequency as the measurand.

Signal conditioning is the overall name for circuits which convert the sensor output to a level suitable for the A/D. They embrace amplifiers, filters and demodulators. All these circuits are described in detail in Peyton and Walsh (1993), Horovitz and Hill (1989) and Arbel (1980). A general text on operational amplifiers which has stood the test of time is by Tobey *et al.* (1971).

Before describing individual circuits, it is worth looking at some more general points.

4.1.1 Earthing or grounding

Most voltages in a system are measured relative to **earth** (or **ground** in American usage), but it is important to remember that not all so-called earth points are at the same potential. Stray capacitance and leakage resistance between points at high potential, such as between mains and earthed surfaces, can cause small currents to flow in earth wires and planes. The resistance of earth wires may be small, but it is not zero, so these currents cause potential drops along the earth wires. Magnetic fields surrounding power transformers and their leads may also induce currents to flow within the supposedly earthed wires. There is always a small voltage drop along a power supply wire or track, so that it is good practice to connect all earth leads to a single point as close to the power supply as possible, and to keep the earth routes for digital and analogue circuits separate. For example, the resistance of the track on one printed circuit board was measured as 3 mΩ cm^{-1}, which is 48 mΩ along the length of a Eurocard, so it is quite possible to have a potential difference of a few millivolts along the earth track. A comprehensive explanation of these problems was given by Brokaw (1984).

4.1.2 Series and common mode noise

It often happens that an extra, unwanted emf appears in series with the emf we are trying to measure. The unwanted emf is called **series mode interference** or, in American usage, **normal mode interference**. Direct current series mode emfs are usually caused by thermoelectric (Seebeck) effects at junctions of dissimilar metals in the circuits, and lead to errors unless the measured voltage can be reversed and measured again. Alternating current series mode noise can be caused by induction from the power lines. Its effect can be greatly reduced by the use of an integrating A/D converter (Chapter 9).

4.1.3 Errors due to common mode interference

Many sensors produce signals at millivolt levels (e.g. strain gauges and thermocouples) and microvolt levels are common in medical work. More often than not, neither side of the signal is earthed. If the signal is truly differential, the common mode voltage is defined as the mean voltage of the two inputs. Thus, one input is $(V_{CM} + \frac{1}{2}V_{sig})$ and the other is $(V_{CM} - \frac{1}{2}V_{sig})$.

Alternatively, the signal may have a 'LOW' side which is approximately at earth potential. Most digital voltmeters are made with two input terminals marked 'HI' and 'LO'. The side of the input which is nearer to earth potential should be connected to LO. In this case the common mode voltage is defined as the voltage from the LO terminal to earth of the voltmeter.

Figure 4.1 shows how a common mode signal can occur in various common measurement circuits. In Figure 4.1(a) one side of the bridge supply voltage is earthed, so the bridge signal is differential, and much smaller than the common mode voltage. This is a common problem.

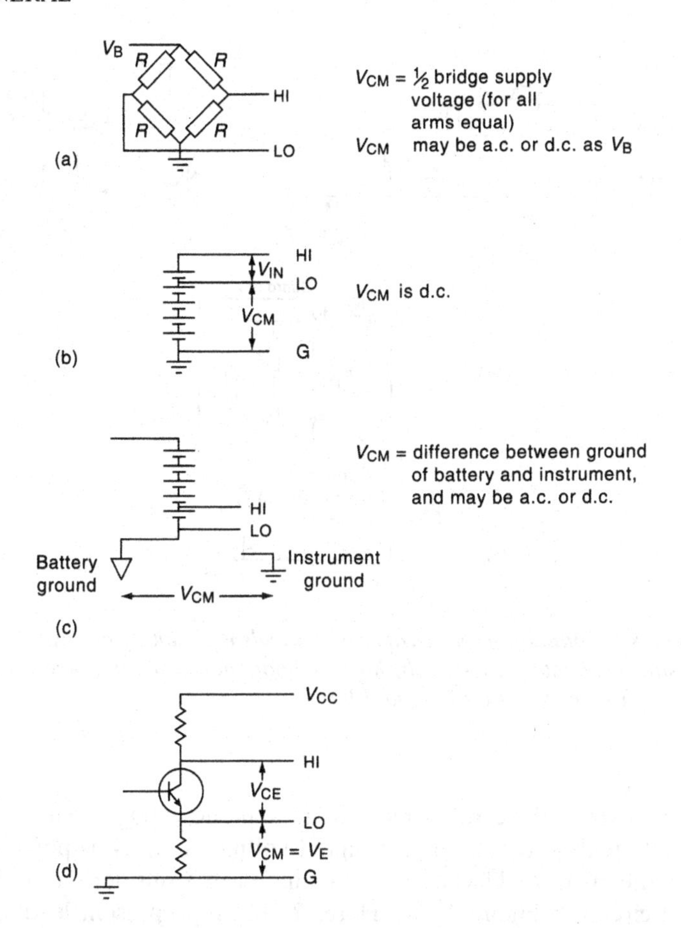

Figure 4.1 *Sources of common mode voltage: (a) bridge; (b) battery; (c) ground loop; (d) transistor.*

In Figure 4.1(b), the negative terminal of a battery is earthed, and a voltmeter is being used to measure the voltage of the most positive cell.

In Figure 4.1(c), the negative side of the measured voltage appears to be at earth potential, but it is unlikely that the instrument earth and the battery earth are at exactly the same potential, as explained above. This is called an **earth** or **ground loop** and can be very troublesome. Figure 4.1(d) shows a digital voltmeter connected to measure the collector–emitter voltage of a transistor. The emitter–earth voltage is in common mode.

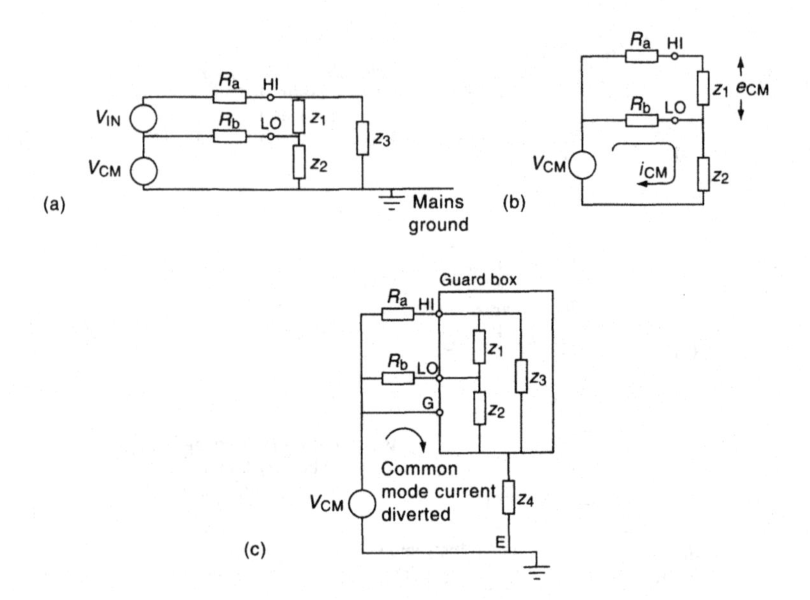

Figure 4.2 *Common mode errors: (a) equivalent circuit of an instrument measuring a floating voltage; (b) how common mode noise becomes series mode noise; (c) the use of a guard box.*

To illustrate the errors which common mode voltages can cause, we will analyse the input to a digital voltmeter which is powered from the mains and has its case at mains earth. Consider the equivalent circuit in Figure 4.2(a). Here, R_a and R_b represent lead and source impedances, z_1 represents the input impedance of the voltmeter, z_2 represents leakage impedance from the LO terminal to mains earth, and z_3 represents leakage impedance from the HI terminal to mains earth.

In a balanced voltmeter $z_2 = z_3$, but more usually in a digital voltmeter $z_3 >> z_2$ and may be ignored.

When $V_{IN} = 0$, the voltage across the HI and LO terminals should be zero. The circuit then reduces to Figure 4.2(b); R_a and z_1 in series together are in parallel with R_b.

Let the impedance of the combination be R_b', then

$$R_b' = \frac{(R_b + z_1)R_b}{z_1 + R_a + R_b} = R_b \text{ (very nearly)}$$

The voltage drop across R_b' is

$$V_b = V_{CM} \frac{R_b'}{R_b' + z_2}$$

The voltage drop across z_1 is

$$e_{CM} = V_{CM} \frac{R_b'}{(R_b' + z_2)} \frac{z_1}{(R_a + z_1)}$$

As $z_2 >> R_b$ and $z_1 >> R_a$, this reduces to

$$e_{CM} = \frac{V_{CM} R_b}{z_2} \tag{4.1}$$

Thus, even with $V_{IN} = 0$, a voltage is produced across the HI and LO terminals by the common mode voltage. If it is d.c. and the meter is on a d.c. range, the common mode voltage will cause a spurious reading and, similarly, if the common mode is a.c. and the meter is on an a.c. range. However, if the common mode is a.c. and the meter is on a d.c. range, the spurious reading will be reduced by a series mode rejection of the integrating A/D converter, as explained in Chapter 9.

4.1.4 Specification of common mode rejection ratio

The performance of a digital voltmeter (dvm) in the presence of a common mode voltage depends on the circuit in which the measurement is made. This is obviously unknown to the dvm manufacturer, so figures are quoted assuming that $R_b = 1000\ \Omega$.

The **common mode rejection ratio** (CMRR) is a figure of merit for digital voltmeters.

The **true CMRR** is defined as

$$20 \log_{10} \frac{V_{CM}}{e_{CM}} \text{ dB} \tag{4.2}$$

This parameter varies with frequency as z_2 includes stray capacitance.

The **effective CMRR** is

$$20 \times \log_{10} \left(\frac{\text{common mode voltage to produce a change of 1 LSD}}{\text{voltage represented by 1 LSD}} \right) \tag{4.3}$$

This definition takes into account the **series mode rejection ratio** (SMRR) of the A/D converter:

$$\text{SMRR} = \frac{\text{maximum noise input}}{\text{maximum error output}}$$

Thus, on a decibel scale,

$$\text{effective CMRR} = \text{true CMRR} + \text{SMRR} \qquad (4.4)$$

4.1.5 The use of a guard terminal

A common mode voltage produces an error reading because of leakage currents flowing from the input terminals to the earthed metal case. If the case were not earthed but kept at the same potential as the LO terminal, there would be no leakage and no error. Although it would not be safe to operate a mains-powered, metal-cased instrument without the case being earthed, we can have two cases, one inside the other. The inner one is called the **guard box** and connected to the **guard terminal**, and the outer case is connected to mains ground.

The user should connect the guard as closely as possible to the source of the common mode voltage as shown in Figure 4.2(c). (This is not always possible, because R_b may represent part of the source impedance of the input and not lead resistance. In this case the guard should be connected to the LO terminal.) The common mode current will then flow through z_4 and avoid R_b. If the input lead is screened, the screen should also be connected to the guard. The guard should not be left unconnected nor connected to ground. Do not confuse G for 'guard' with G for 'ground' (British notation is easier here because E for earth is used in place of G for ground).

The guard box surrounds the analogue circuits but not the digital circuits in a dvm. If common mode errors are much less than 1 LSD, a guard is an unnecessary complication and expense, so they are generally used only on metal-cased voltmeters of 5½ digits resolution or more. Figure 4.3 shows the arrangement of earth connections in such an instrument. Note that all earth wires are connected to a single point close to the power supply. They should **not** be connected like tributaries into a river. A good description of the principles of guard circuits will be found in Hewlett-Packard Application Note Number 123, Hewlett-Packard (1989).

4.2 Instrumentation amplifiers

Whatever else may be required, the signal conditioning will almost always need an amplifier. The preferred type for most applica-

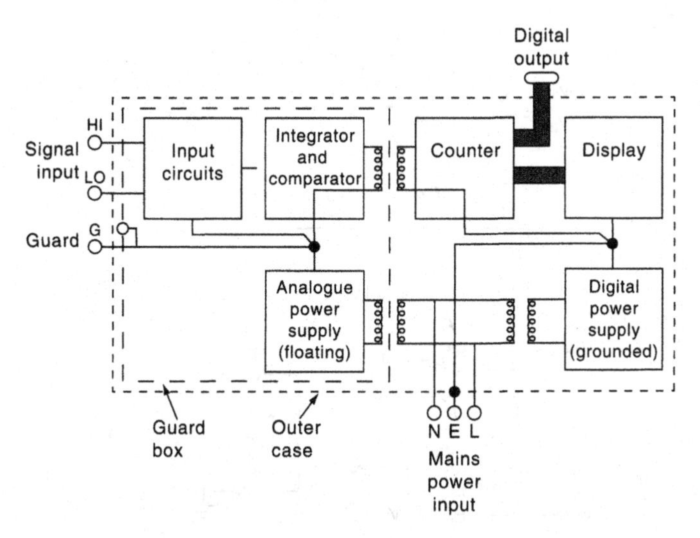

Figure 4.3 *Ground connections in a high-sensitivity instrument.*

tions is an **instrumentation amplifier**. This name is applied to operational amplifier circuits which accept a differential input signal and have a single-sided output. In a simple differential amplifier such as that shown in Figure 4.4(a) where $R_1 = R_1'$ and $R_2 = R_2'$, the differential gain is R_2/R_1 and the differential input impedance is $2R_1$. The CMRR depends on how closely the resistors are matched and on the CMRR of the op-amp device itself. The term instrumentation amplifier is used occasionally to mean a high-quality, single operational amplifier, but is more correctly applied to the circuit in Figure 4.4(b), made up of three operational amplifiers as shown. Because both inputs are connected to the noninverting inputs of the op-amps, the impedance is high to both inverting and noninverting inputs. This circuit is available as a single IC from many manufacturers.

4.2.1 Differential gain

Referring to Figure 4.4(b), the open-loop gain of each op-amp is very high, so $e_1 = V_1$ and $e_2 = V_2$. Therefore, the voltage across R_4 is given by

$$e_1 - e_2 = V_1 - V_2$$

and the current through R_4 is given by

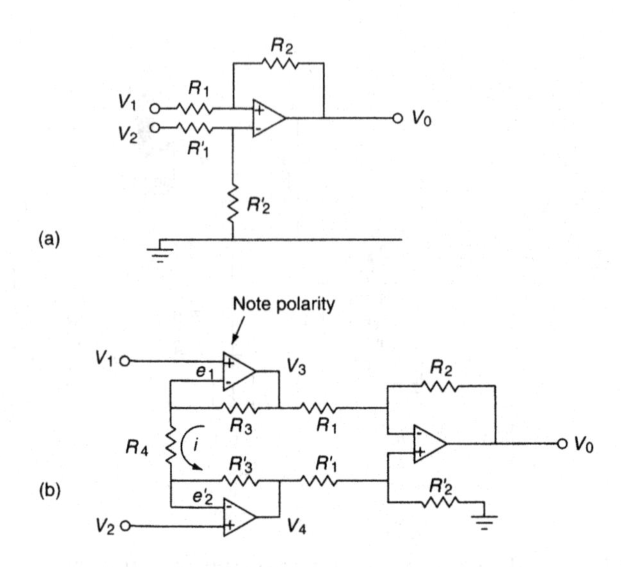

Figure 4.4 *Instrumentation amplifiers: (a) single op-amp; (b) triple op-amp.*

$$i = \frac{V_1 - V_2}{R_4}$$

This flows through R_3, R_4 and R_1 as shown, but is not drawn from the input.

The second stage is a conventional differential amplifier. The differential input to the second stage is

$$V_3 - V_4 = i(R_3 + R_4 + R_3')$$

$$= \frac{V_1 - V_2}{R_4}(R_3 + R_4 + R_3')$$

Now $R_1 = R_1'$, $R_2 = R_2'$ and $R_3 = R_3'$, and the differential gain of second stage is R_2/R_1.

Therefore, the overall gain is given by

$$\frac{R_1}{R_2} \cdot \frac{R_4 + 2R_3}{R_4} \tag{4.5}$$

The component R_4 is the **gain resistor** and, in integrated instrumentation amplifiers, it can be adjusted by the user; the values of the other resistors are fixed by the manufacturer.

4.2.2 Common mode gain

Referring again to Figure 4.4(b), let the common mode voltage $= V_1 = V_2 = V_{CM}$. Then, $i = 0$ and $V_3 = V_4 = V_{CM}$. That is, the first stage amplifies the differential signal, but not the common mode signal. In practice, the resistors cannot be perfectly matched and the common mode signal is not completely eliminated. The overall CMRR is

$$20 \log_{10}\left(\frac{\text{differential gain}}{\text{common mode gain}}\right) \tag{4.6}$$

In choosing a suitable amplifier for a particular application, the specifications to note are the gain accuracy and the nonlinearity, the input offset voltage and input offset bias current, and also the temperature coefficients of all these parameters. For dynamic signals, the bandwidth and the maximum slew rate must also be examined. The slew rate is the maximum rate of change of voltage and is a function of both frequency and amplitude. Bandwidth varies with gain, being larger at low gains. In general, small offsets tend to be associated with lower slew rates.

4.3 Isolation amplifiers

If the difference between the low potential of the signal source and the earth of the A/D converter is too large to be accommodated by the CMRR of an instrumentation amplifier, it is necessary to provide isolation. No matter how high the CMRR of an instrumentation amplifier, both input and output signals must lie between the voltages of the power supplies, usually ± 15 V with respect to earth.

Even when the sensor is near earth potential, isolators are used in medical applications to protect the patient from potentially dangerous voltages which might occur due to a fault in the equipment. This illustrates the design principle that even if the probability of a fault is very small, but the consequence is very serious, the design must be modified so that if the fault does occur, then the disastrous consequence will not follow.

An A/D converter may be isolated as shown in Figure 4.3. The first part of the converter generates a train of pulses which pass through an isolating transformer or an electro-optical isolator before they are counted. Alternatively, an **isolation amplifier** or **isolator** may be incorporated in the signal conditioning.

An isolation amplifier comprises a differential input amplifier stage, driving a modulator. Pulses are transferred across the isolation barrier by a small transformer. On the output side, the pulses are demodulated and filtered. The signal is usually sent to the output via a unity-gain buffer. Obviously, the power supplies for input and output amplifiers must be separate and isolated. In a three-port isolator, the input and output power supplies are driven from a master power oscillator through isolating transformers, but in a two-port isolator the power supply unit is isolated from the input but not the output stage. In a multichannel system, all channels may be driven from a single oscillator to eliminate the risk of noise being introduced at the beat frequency of neighbouring oscillators.

In place of pulse transformers, some isolators employ modulation and coupling through a very small capacitor, and others use opto-isolators without modulation. Peyton and Walsh (1993) describe a technique using feedback to compensate for the nonlinearity of the opto-isolator.

Isolators are readily available which can stand 2000 V d.c. (or peak a.c.) between input and output, and which have CMRRs of 110 dB or more at 60 Hz (e.g. AD202, 204 and 208, and Burr–Brown ISO 102, 120, 121 and 122). Models are also available which incorporate a guard box as described above. The modulation–demodulation process restricts the bandwidth, but as the carrier frequency is of the order of 50 kHz or higher, the bandwidth may extend up to tens of kHz.

4.4 Charge amplifiers

4.4.1 Requirements

The output from a piezoelectric sensor is in the form of voltage or charge, and the internal resistance of the sensor is very large; $10^8\ \Omega$ is not unusual. A correspondingly high impedance input amplifier must be used, i.e. a field effect transistor (FET) input stage. Any bias current from the amplifier input will charge the capacitance of the sensor and cause drift. It is therefore necessary to provide either a reset switch or a high-resistance path for the bias current, typically $10^9\ \Omega$.

A voltage amplifier has the disadvantage that the overall sensitivity depends on the capacitance of the sensor, the connecting cable, strays and the amplifier input, as charge developed by the sensor is shared amongst all the capacitances. Thus if a sensor has a new connecting cable fitted, the calibration is changed. Pity the poor engineer! A **charge amplifier** is an operational amplifier circuit whose output voltage is proportional to the charge applied at the input, and it overcomes the problem as described in the next section.

4.4.2 Circuit

Figure 4.5 shows the simplified circuit of a charge amplifier. The sensor is connected directly to the input of the op-amp with no input resistor. This point is a virtual earth in the usual way and all charge developed by the sensor is transferred to the feedback capacitor C_f. Because there is negligible voltage across the cable and stray capacitances C_1, they have negligible effect. The output voltage is given by

$$V_{out} = \frac{Q_{in}}{C_f} \tag{4.7}$$

The manufacturers of piezoelectric accelerometers and force transducers quote the sensitivity both in terms of mV g^{-1} or mV N^{-1} for use with high input impedance voltage amplifiers, and in pC g^{-1} or pC N^{-1} for use with charge amplifiers.

A reset switch and a large resistor for leakage currents may be connected across the capacitor. A resistor is usually necessary with ceramic piezo sensors, but not with quartz ones.

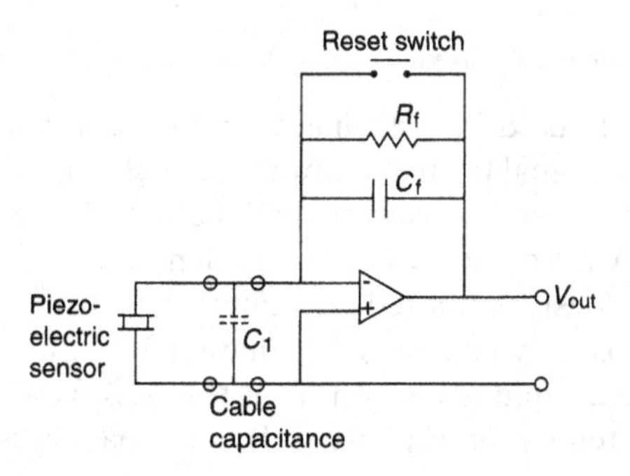

Figure 4.5 *Charge amplifier.*

4.4.3 Frequency response

In Figure 4.5, the inverting input is a virtual earth so that no current flows into C_1. If the force on the crystal is alternating, as when measuring a vibration, then a.c. flows from the crystal into the parallel combination of R_f and C_f:

$$i = V_0\left(\frac{1}{R_f} + j\omega C_f\right) = j\omega Q \qquad (4.8)$$

$$V_0 = j\omega Q \frac{R_f}{1 + j\omega R_f C_f} \qquad (4.9)$$

$$V_0 \text{ approaches } \frac{Q}{C_f} \quad \text{when} \quad j\omega R_f C_f >> 1 \qquad (4.10)$$

This is the normal operating condition. Also,

$$|V_0| = \frac{1}{\sqrt{2}} \cdot \frac{Q}{C_f} \quad \text{when} \quad \omega = \frac{1}{R_f C_f} \qquad (4.11)$$

The output is 3 dB below the midfrequency value; this is the lower limit of working frequency. The response to a step-function change of force or acceleration is a corresponding step function change in output followed by an exponential return to the original level with the time constant R_fC_f. Thus R_f should be as large as possible without the output drifting when the input is constant.

4.5 Filters

4.5.1 Classification of filters

Filters are widely used in instrumentation to separate the wanted components of a signal from the unwanted ones. For example, some sensors modulate an a.c. carrier, and before the signal can be applied to an A/D converter it must be demodulated and the carrier frequency must be filtered out. Filters are also used to avoid aliasing, which is fully explained in Chapter 10. Filter applications usually require a rapid transition from the pass-band to the stop-band, but no attenuation or phase shift in the pass-band and 100% attenuation in the stop-band. These requirements are mutually contradictory, and it is the designer's task to find the best compromise for any particular application. **Passive filters** comprise a network of capacitors and inductors or resistors; **active filters** include an operational amplifier. Active filters are preferred for instrumentation applications because their characteristics are more easily controlled and they avoid the bulk and nonlinearity of inductors.

The transfer function is defined as

$$H(\omega) = \frac{\text{output voltage}}{\text{input voltage}} \tag{4.12}$$

and is a complex function of frequency. In the stop-band, the output voltage of a single-pole, low-pass filter is inversely proportional to frequency, and the phase shift is 90°. A more precise way of saying the same thing is that the attenuation is 6 dB per octave = 20 dB per decade. (The term 'octave' is used here to mean twice the frequency, by analogy with musical terminology.) A filter with N poles has a stop-band attenuation of $20 \times N$ dB per decade and a phase shift of $90 \times N°$. In an active filter each op-amp contributes two poles.

Filters may be classified in several ways. First, the frequency response may be low-pass, high-pass, band-pass or notch. This is illustrated in Figure 4.6 where the solid lines indicate the idealized response and the dotted lines the practical response. In each case, an ideal filter would not introduce any phase shift but a real one always will.

Second, real filters may be classified according to the way the frequency and phase responses differ from perfection. The three types which are discussed here are Butterworth, Bessel and Chebyshev.

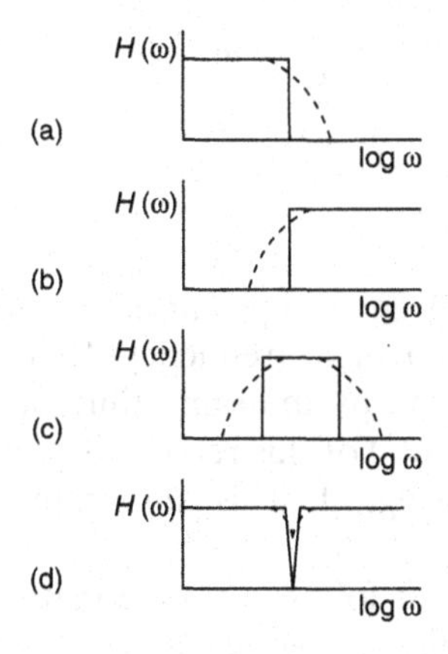

Figure 4.6 *Idealized filter responses: (a) low-pass; (b) high-pass; (c) band-pass; (d) notch.*

Finally, active filters may be classified by the type of circuit used to realize the response. A few examples are given below, but more detailed treatments are given by Peyton and Walsh (1993), Horovitz and Hill (1989), Tobey *et al.* (1971) or Bowron and Stephenson (1979). Digital filters are not discussed here as they operate on digitized data and we are now considering the analogue circuits before the A/D converter.

4.5.2 Types of filter response

Butterworth response

This type of filter has the least attenuation in the pass-band, hence its alternative name of 'maximally flat' (Figure 4.7). The trade-off for the good amplitude response is phase shift at frequencies well below the break frequency. The break frequency of a Butterworth filter is the frequency where the amplitude is reduced to 70.7% of its value in the pass-band, i.e. the attenuation is –3 dB and the phase shift is 90° per pole.

The amplitude transfer function of a low-pass, Butterworth filter with N poles is

$$|H(\omega)| = \frac{H(0)}{\sqrt{(1 + \Omega^{2N})}} \tag{4.13}$$

where Ω is the **normalized frequency**, that is, the ratio of the actual frequency to the break frequency:

$$\Omega = \frac{\omega}{\omega_0} \tag{4.14}$$

Bessel response

This type of filter is designed to produce a phase shift in the pass-band which is proportional to frequency. That is, all components of the signal are delayed by the same time, so that the signal is not distorted by the filter. For this reason, Bessel filters are recommended for anti-aliasing, if it is important to preserve the waveform.

The response does not follow a simple mathematical equation, but is shown graphically in Figure 4.8. In this type of filter, the normalized response is shown for filters which delay all signals within the pass-band by one second. Thus if the delay is only

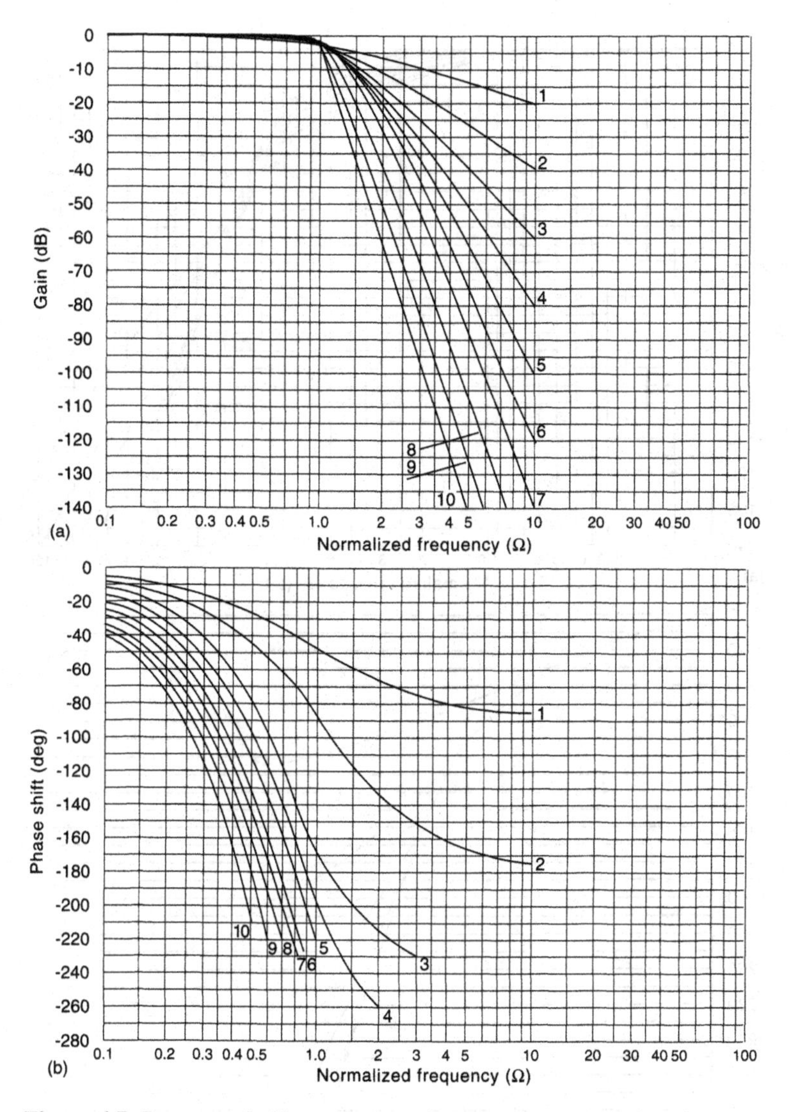

Figure 4.7 *Butterworth filter response: (a) gain–frequency; (b) phase–frequency. The number of poles is given against each curve.*

0.1 s, the attenuations shown in Figure 4.8 at 1 rad s^{-1} will occur at 10 rad s^{-1}. One second corresponds to a phase shift of 360° for a frequency of 1 Hz or 1 rad (57.3°) for a frequency of 1 rad s^{-1}. The phase–frequency graph is curved because frequency is plotted on a logarithmic scale.

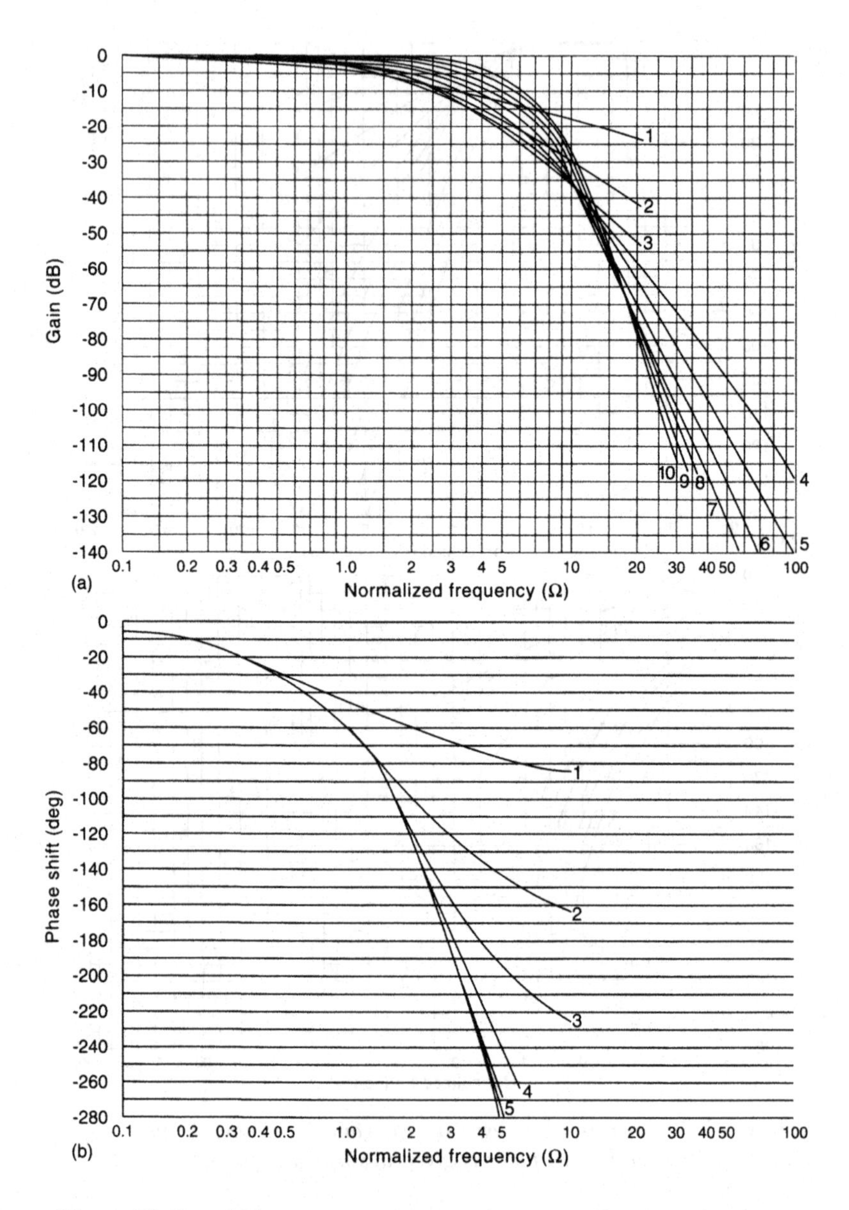

Figure 4.8 *Bessel filter response: (a) gain–frequency; (b) phase–frequency. The number of poles is given against each curve.*

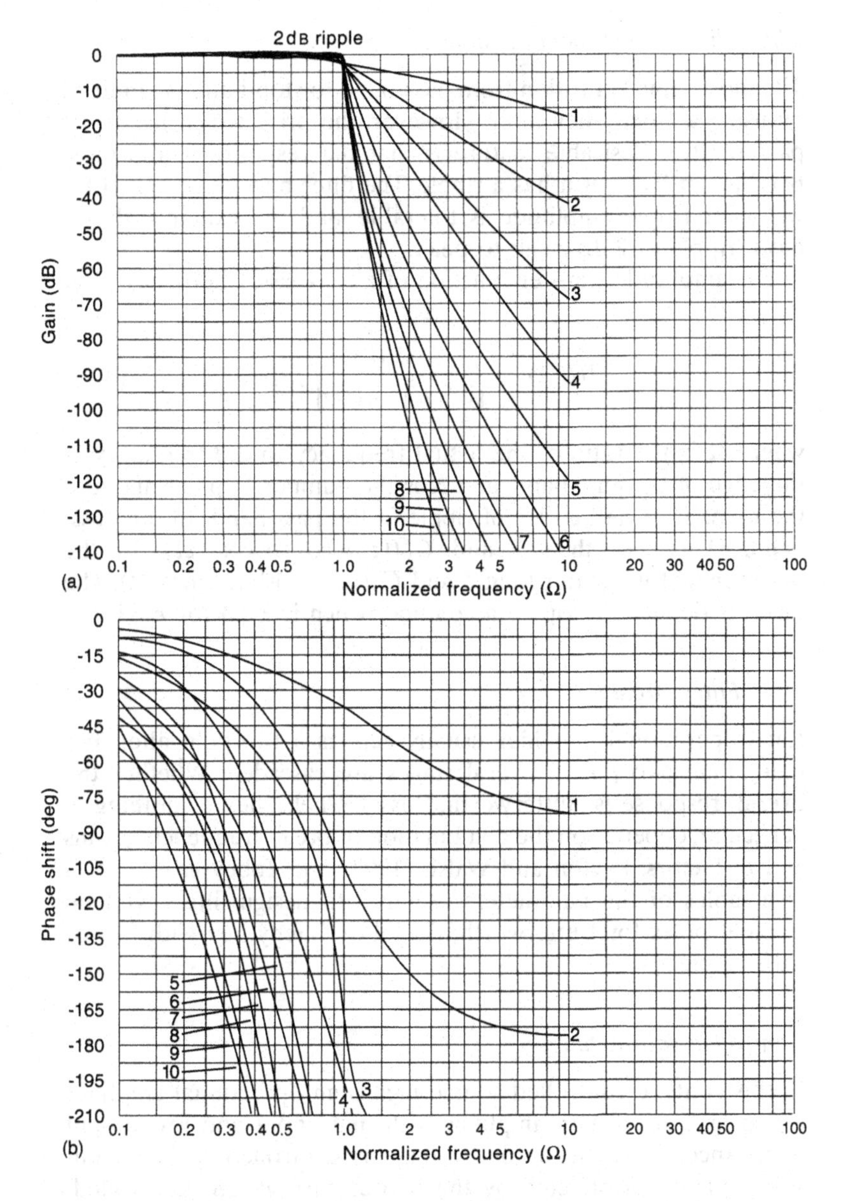

Figure 4.9 *Chebyshev filter response: (a) gain–frequency; (b) phase–frequency. The number of poles is given against each curve.*

Chebyshev (Tchebyshev) responses

The most rapid transition from pass-band to stop-band is obtained by using a Chebyshev filter. However, the price which has to be paid for this desirable feature is that the response in the pass-band is not flat, but it has a ripple. The filter may be designed for whatever ripple amplitude is tolerable, and the example shown has a ripple of 2 dB peak to peak.

The amplitude transfer function of a low-pass, Chebyshev filter is

$$|H(\Omega)| = \frac{H(0)}{\{1 + \varepsilon^2[C_N(\Omega)]^2\}^{1/2}} \tag{4.15}$$

where Ω is the ratio of the actual frequency to the frequency at the edge of the pass-band, and N is the number of poles and also the number of half-cycles of ripple in the pass-band. In the pass-band, Ω is less than 1 and $C_N(\Omega) = \cos(N \cos^{-1} \Omega)$. In the stop-band, Ω is greater than 1 and $C_N(\Omega) = \cosh(N \cos^{-1} \Omega)$. The factor ε defines the ripple amplitude which is 3 dB for $\varepsilon = 1$.

4.5.3 Filter circuits

Each operational amplifier usually acts as a two-pole filter. For more than two poles, several stages are cascaded. Whether the overall response is Butterworth, Bessel, Chebyshev or nothing in particular, depends on the combination of break frequencies, gains and Q factors. Peyton and Walsh (1993) and Tobey *et al.* (1971) give tables of the parameters required to design filters with up to eight poles for Butterworth, Bessel or Chebyshev with ½, 1 or 3 dB ripple.

Controlled-source filters

In this context, controlled source means an operational amplifier acting as a fixed gain amplifier with high input and low output impedance. Thus the output voltage is controlled by the input voltage and not affected by the output current. The controlled-source circuit is shown in Figure 4.10(a), and is denoted by the symbol K inside a triangle in Figure 4.10(b).

The transfer function of the circuit in Figure 4.10(b) is

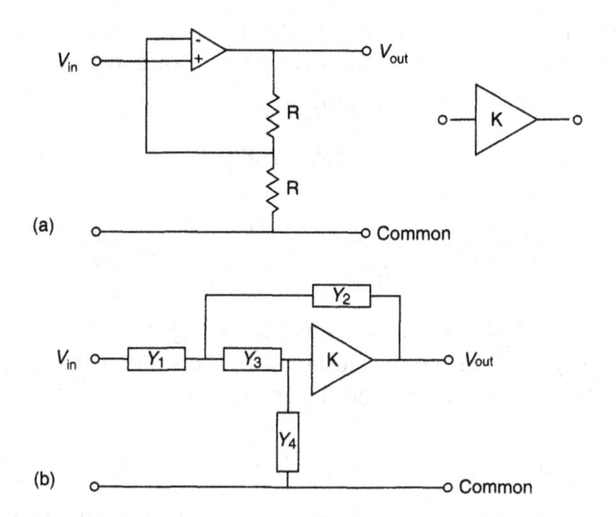

Figure 4.10 *Controlled-source filter.*

$$\frac{V_{out}}{V_{in}} = |H(\omega)|$$

$$= \frac{KY_1Y_2}{Y_4(Y_1 + Y_2 + Y_3) + Y_3[Y_1 + Y_2(1 - K)]} \tag{4.16}$$

To make a low-pass filter, Y_1 and Y_3 become resistors, and Y_2 and Y_4 capacitors:

$$Y_1 = \frac{1}{R_1}; \quad Y_3 = \frac{1}{R_2}; \quad Y_2 = j\omega C_1; \quad Y_4 = j\omega C_2$$

Substituting gives

$$|H(\omega)| = \frac{K}{1+j\omega[R_2C_2+R_1C_2+R_1C_1(1-K)]+(j\omega)^2R_1C_1R_2C_2} \tag{4.17}$$

or, in Laplace notation,

$$|H(s)| = \frac{|H(0)|}{1+s[R_2C_2+R_1C_2+R_1C_1(1-K)]+s^2R_1C_1R_2C_2} \tag{4.18}$$

$$|H(s)| = \frac{|H(0)|}{1 + (s/Q\omega_0) + (s^2/\omega_0^2)} \tag{4.19}$$

By comparing equations we can find the parameters of the filter response, namely $|H(0)|$, ω_0 and Q in terms of component values:

$$|H(0)| = K \tag{4.20}$$

$$\omega_0^2 = (R_1C_1R_2C_2)^{-1} \tag{4.21}$$

$$\frac{1}{Q} = \left(\frac{R_2C_2}{R_1C_1}\right)^{1/2} + \left(\frac{R_1C_2}{R_2C_1}\right)^{1/2} + \left(\frac{R_1C_1}{R_2C_2}\right)^{1/2} - K\left(\frac{R_1C_1}{R_2C_2}\right)^{1/2} \tag{4.22}$$

If $R_1 = R_2$ and $C_1 = C_2$, this simplifies to $1/Q = 3 - K$.

A high-pass filter can be designed in the same way, but Y_1 and Y_2 are capacitors and Y_2 and Y_4 are resistors.

State-variable filters

This is a particularly versatile design which is available in integrated form from several manufacturers, described as a Universal Active Filter, such as the Burr–Brown UAF42. It consists of two active integrators and one amplifier. (An extra amplifier is sometimes available in the same package.) Low-pass, high-pass and band-pass outputs are available at the same time from different points in the circuit. A typical circuit is shown in Figure 4.11. The four resistors enclosed in rectangles are externally connected and the manufacturers provide details, including a computer program, of how to calculate the values to obtain different responses (Burr–Brown AB-035).

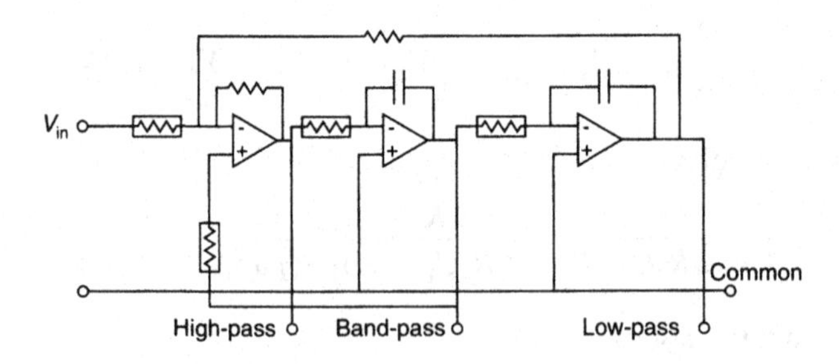

Figure 4.11 *State-variable filter.*

Switched-capacitor filters

A capacitor which is alternately charged from a source and discharged elsewhere behaves like a resistor whose value is controlled by the switching frequency. The current drawn from the source is unidirectional. For example, a capacitor C, is charged to a voltage V and discharged, and this is repeated at a frequency f. The charge transferred in each period is

$$q = CV$$

The charge transferred in each second, i.e. the current, is

$$qf = i = CVf$$

The effective resistance is given by

$$\frac{V}{i} = \frac{1}{fC} \tag{4.23}$$

Note that when the capacitor is used in the usual way, charge is returned to the source, and

$$\frac{V}{i} = \frac{1}{2\pi fC}$$

This current is alternating.

There are a number of versatile filter ICs on the market using this principle. The best known is probably the MF10 from National Semiconductor which contains two, two-pole filter sections which can be used separately, or cascaded to make a four-pole filter. Each section can be configured as a low-pass, high-pass, band-pass or notch function with any of the responses described above. The great advantage of a switched-capacitor filter is that its cut-off frequency is proportional to the switching frequency, and is, therefore, easily controllable. The switching frequency is always higher than the cut-off frequency, usually 50 or 100 times higher. The main disadvantage is that the switching may generate noise on the signal.

4.6 Integrators and differentiators

If $v = V \sin \omega t$ then

$$\int v \, dt = \frac{-V \cos \omega t}{\omega}$$

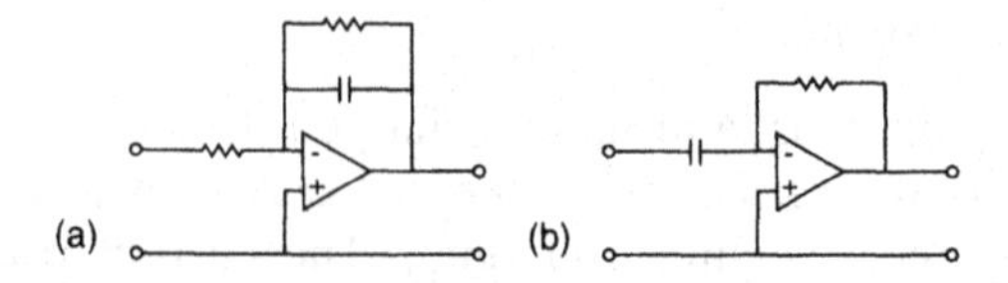

Figure 4.12 *(a) Integrator and (b) differentiator.*

and $dv/dt = \omega V \cos \omega t$. Thus an integrator is a single-pole, low-pass filter operated above the break point, and a differentiator is a single-pole, high-pass filter operated below the break point. Typical circuits are shown in Figure 4.12.

Integrators are prone to drift because any bias current from the op-amp will flow into the capacitor and charge it steadily. For this reason a high-value resistor is connected across the capacitor. The time-constant of this resistor and the integrating capacitor must be long compared with the period of the signal to be integrated. Sometimes a reset switch is added so that the capacitor can be discharged.

Although differentiating circuits do not suffer from drift, they do suffer from noise because high frequency signals are amplified. It is therefore necessary to restrict the bandwidth of the differentiator and set the break point just beyond the working range.

4.7 Phase-sensitive detectors

4.7.1 Applications

Some sensors, such as linear variable differential transformers (LVDTs), modulate an a.c. carrier, so the signal must be 'de-modulated' or 'detected' before it can be applied to an A/D converter. (The term 'detected' comes from the early days of radio when the audio signal was said to be detected in the modulated carrier.) The simplest detector is a rectifier, but this has the serious disadvantage that displacements either side of the null position produce exactly the same rectified output. This type of detector produces a signal in phase with the carrier for displacement in one direction and 180° out of phase for displacement in the other. A phase-sensitive detector (PSD) compares the signal with the carrier and produces a positive output for displacement in one direction and negative in the other. It has the further advantage that it produces no output for input components at frequencies other than the carrier, such as mains interference, neither does it produce any output for inputs in quadrature with the carrier, which can be caused by stray capacitance in the sensor circuit.

A PSD requires two inputs, the signal and the reference, and is followed by a low-pass filter. In a linear type, the output is proportional to the signal input and is a function of the phase angle between it and the reference. Provided that the reference is more than a threshold value, its magnitude does not affect the output, because it is first converted to a constant amplitude square wave and then used either to control a switch and reverse the direction of the signal.

Alternatively, the squared reference and the signal may be fed into an analogue multiplier. An exclusive-OR circuit may be regarded as a completely digital PSD.

4.7.2 Linear, switching PSD

Figure 4.13 shows a circuit for a PSD and Figure 4.14 shows the waveforms for phase angles of 0°, 90° and 180° between the signal and the reference. When they are the same frequency and in phase, the output is a full-wave rectified sine wave as in Figure 4.14(a). The low-pass filter removes components of the reference frequency and above and leaves a positive d.c. output. When the signal and reference are in antiphase, the d.c. output is negative as in Figure 4.14(b). When the signal and reference are in quadrature, there is no d.c. output. This can be very useful in conditioning the signal from a sensor in a bridge circuit as quadrature signals can be produced by stray capacitance, whereas the true signal is in phase with the reference.

In Figure 4.13(a), the field effect transistor F_1 is switched on and off by the reference signal. When it is OFF, both input terminals of the op-amp follow the signal and the gain is + 1, as in Figure 4.13(b), but when it is ON, the noninverting input of the op-amp is held almost to earth and the gain is almost –1 (Figure 4.13(c)). The small fraction of the input which reaches the noninverting input with F_1 ON can be compensated by making the resistor to the noninverting input twice the others and putting a similar FET, F_2 which is always ON, in the feedback. This is demonstrated in Example 4.2.

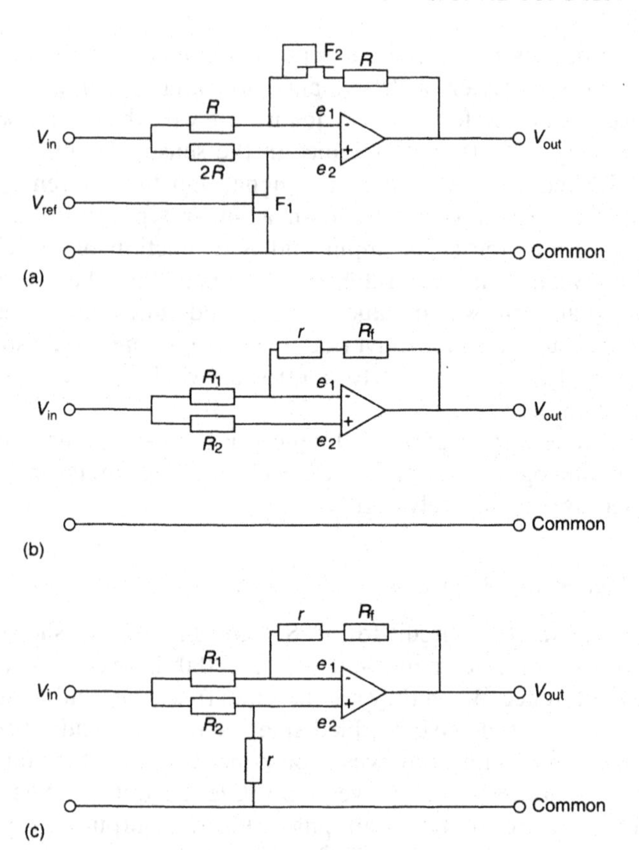

Figure 4.13 *Phase-sensitive detector: (a) circuit; (b) equivalent when F_l is OFF; (c) equivalent when F_l is ON.*

Analysis of a linear, switching PSD

Let the signal be $A \sin(\omega t + \phi)$ and the gain be switched between $+k$ and $-k$ at the reference frequency so that

$$\text{gain} = +k \qquad \text{when } 0 < t < \pi/\omega_r$$

$$\text{gain} = -\text{k} \qquad \text{when } \pi/\omega_r < t < 2\pi/\omega_r$$

Although the use of angular frequency is not very meaningful for a square wave, the use of ω_r instead of $2\pi f_r$ simplifies the equations.

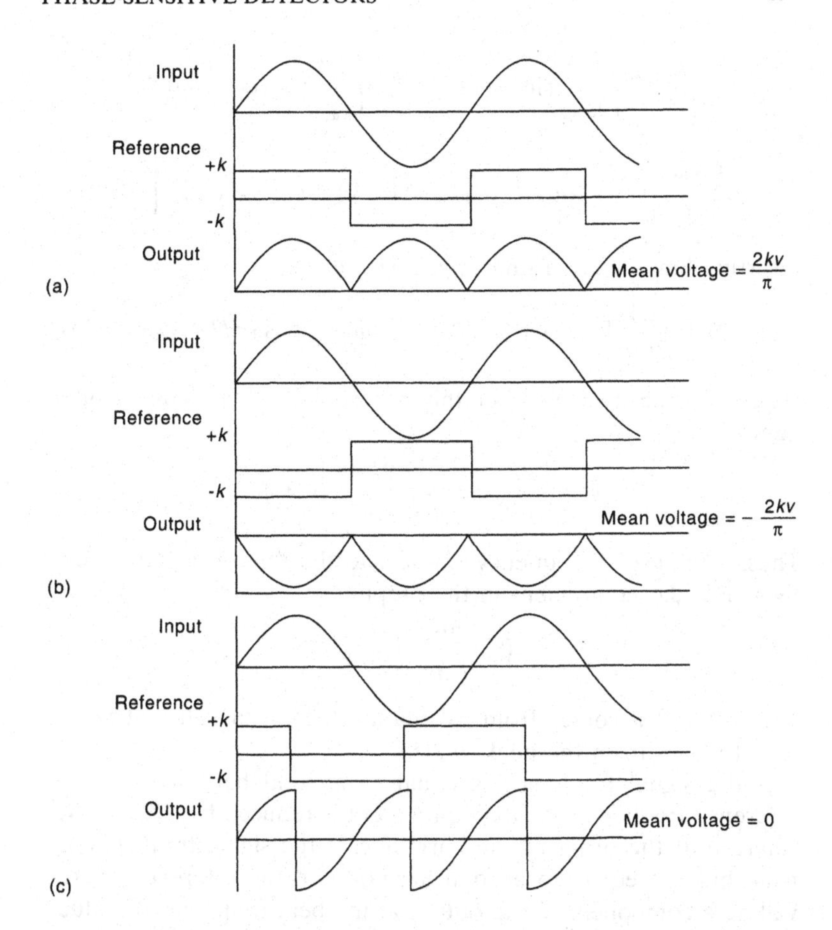

Figure 4.14 *Phase-sensitive detector waveforms: (a) signal in phase with reference; (b) signal 180° out of phase with reference; (c) signal lags reference by 90°.*

Let the output voltage be V, then

$$V = kA \sin(\omega_s t + \phi), \text{ from } \omega_r t = 0 \text{ to } \omega_r t = \pi$$

and

$$V = -kA \sin(\omega_s t + \phi), \text{ from } \omega_r t = \pi \text{ to } \omega_r t = 2\pi$$

Then, the mean value of the output over one period of the reference input is

$$\overline{V} = \frac{kA\omega_r}{2\pi}\left[\int_0^{\pi/\omega_r} \sin(\omega_s t+\phi)\,dt - \int_{\pi/\omega_r}^{2\pi/\omega_r} \sin(\omega_s t+\phi)\,dt\right]$$

$$= \frac{kA\omega_r}{2\pi\omega_s}\left\{[-\cos(\omega_s t+\phi)]_0^{\pi/\omega_r} - [-\cos(\omega_s t+\phi)]_{\pi/\omega_r}^{2\pi/\omega_r}\right\}$$

Inserting limits and writing N for ω_s/ω_r gives

$$\overline{V} = \frac{kA}{2N\pi}[\cos\phi + \cos(2\pi N+\phi) - 2\cos(\pi N+\phi)] \quad (4.24)$$

When N is an even number this is zero, but when N is an odd number

$$\overline{V} = \frac{2kA}{N\pi}\cos\phi = \frac{2kAf_r}{\pi f_s}\cos\phi \quad (4.25)$$

Thus, if the signal frequency is equal to the reference frequency, there is a d.c. component in the output:

$$\overline{V} = \frac{2kA}{\pi}\cos\phi$$

The factor $2/\pi$ comes from calculating the mean value of sinusoidal pulses from the peak value.

If the signal frequency is equal to an odd harmonic of the reference frequency, the d.c. component is reduced by a factor N, where N is the order of the harmonic. If the signal frequency is near, but not equal, to an odd harmonic of the reference, there will be a component in the output at the beat frequency of value $N \times f_r \pm f_s$, provided this is low enough to pass through the low-pass filter which follows the PSD. Thus the PSD and its low-pass filter together behave as a tuned filter, passing only the reference and its odd harmonics. The sharpness of the tuning depends on the break point of the low-pass filter.

4.7.3 Multiplying PSD

The same effect as above may be achieved by applying the signal and reference voltages to an analogue multiplier circuit.

An alternative method of analysing either version of the circuit is to consider the reference signal (which is a square wave) as a Fourier series and to multiply it by the signal, namely

$$A\sin(\omega t+\phi)$$

$$V_{\text{ref}} = \frac{4k}{\pi}\left(\sin \omega t - \frac{\sin 3\omega t}{3} + \frac{\sin 5\omega t}{5} - \frac{\sin 7\omega t}{7} \ldots\right) \tag{4.26}$$

4.7.4 Digital PSD

Two pulse trains are applied to the inputs of an exclusive-OR circuit, so that the output with be HIGH when one input is HIGH and the other is LOW or vice versa. If both signals are the same frequency, the output is zero when the phase difference is 0° and maximum when it is 180°. As with linear PSDs, the output pulse train is low-pass filtered. Unlike linear circuits, the filtered output is directly proportional to ϕ, not to $\cos \phi$, but only in the range from 0° to 180°.

4.7.5 Edge-triggered PSD

This is another type of digital PSD, sometimes known as type 2. Both the reference and the signal are square waves, and set a bistable on the falling edge of the reference and reset it on the falling edge of the signal. Some additional circuitry converts the bistable output to positive pulses when the signal lags the reference and negative pulses when it leads. When the two are in phase, there are no output pulses at all.

4.7.6 Phase-locked loop

This versatile circuit is made by using the output of a PSD to drive the frequency control of a voltage-controlled oscillator (VCO). The oscillator provides the reference voltage to the PSD either directly or after a frequency divider (Figure 4.15). Provided that the input frequency is within the lock range (see below), the oscillator will lock its frequency to the input frequency, and the circuit will behave as a sharply tuned filter passing only the signal frequency and its odd harmonics. Phase-locked loops (PLLs) have many applications, including:

- extracting very small signals from a noisy background;
- generating a sinusoidal signal at the same frequency as a noisy, square or triangular signal;
- filtering a variable frequency signal with the pass-band tracking the signal frequency;

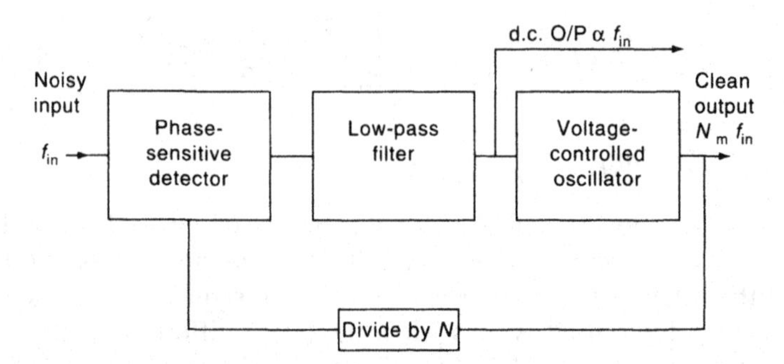

Figure 4.15 *Phase-locked loop.*

- demodulating a frequency-modulated signal;
- with a frequency divider in the feedback path, generating a frequency which is a multiple of the signal frequency.

The **lock range** is the range of frequencies which the PLL will follow once it has locked. It is determined by the range of the control voltage, $\pm \Delta V_C$, and the sensitivity of the VCO, df/dV_C. Thus,

V_C can vary from $\quad V_C$ to $V_C \pm \Delta V_C$

f_{osc} can vary from $\quad f_0$ to $f_0 \pm \Delta V_C\, (df/dV_C)$

The lock range $2f_L$ is given by

$$2f_L = 2\,\Delta V_C \frac{df}{dV_C} \tag{4.27}$$

If there is *a* divided by N in the feedback, substitute (f_{osc}/N) for f_{osc}. A PLL with a single pole filter may not always lock, but it will always be stable.

The **capture range** is the range of frequencies to which the PLL will lock after running free, and is less than the lock range. The detailed analysis of the circuit depends on the type of PSD (linear or digital), the characteristics of the filter (number of poles, gain and breakpoints(s)) and whether there is any additional gain in the loop. Analyses of several circuits are given by Connelly (1973), Gardner (1979), Meade (1982) and Horowitz and Hill (1989).

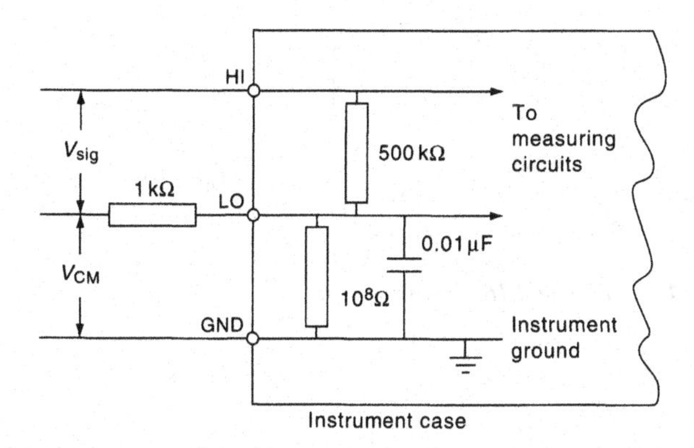

Figure 4.16 *Input circuit of a dvm.*

4.8 Examples

Example 4.1

Common mode interference can affect any mains-powered instrument. The next two examples illustrate the design of front-end circuits to minimize the error.

Figure 4.16 shows the equivalent input circuits of a dvm. Calculate the true CMRR at d.c. and estimate it at 1592 Hz using reasonable approximations.

Solution

Consider the circuit when $V_S = 0$. V_{CM} is effectively connected to the HI and LO terminals together and earth, as in Figure 4.17. The resistance of 1 kΩ and 500 kΩ in parallel is

$$\frac{1 \times 500}{501} = 0.998 \text{ k}\Omega = 998\ \Omega$$

At d.c., the voltage across the HI and LO terminals is given by

$$V_{HL} = V_{CM}\frac{998}{998 + 10^8}$$

$$= V_{CM} \times 9.98 \times 10^{-6}$$

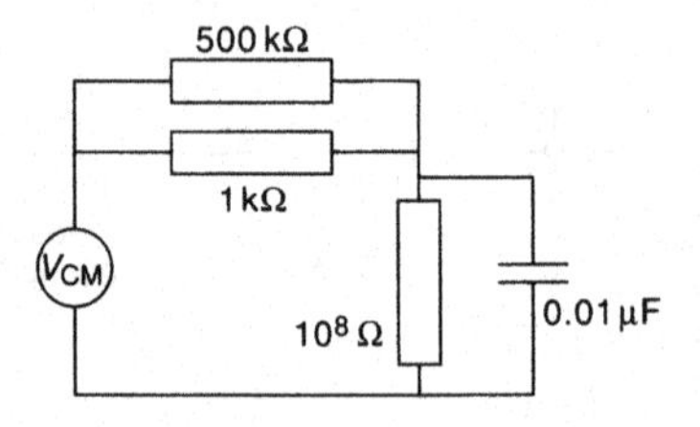

Figure 4.17 *As Figure 4.16, when* $V_{sig} = 0$.

Therefore

$$\text{CMRR (d.c.)} = \frac{V_{CM}}{V_{HL}} = \frac{1}{9.98 \times 10^{-6}} = 100\,200 = 100.02 \text{ dB}$$

At 1592 Hz, $2\pi f = 10^4$. Reactance of 0.01 μF capacitor is

$$\frac{-j}{\omega C} = \frac{-j}{10^4 \times 10^{-6} \times 0.01} = -j \times 10\,000\ \Omega$$

The impedance from LO to earth Z_{LE} is 100 MΩ and 0.01 μF in parallel is

$$Z_{LE} = \frac{10^8(-j \times 10^4)}{10^8 - j \times 10^4}$$

As the question asks us to 'estimate' the a.c. CMRR, we are justified in approximating this to

$$Z_{LE} = \frac{10^8\,(-j \times 10\,000)}{10^8} = -j \times 10\,000$$

and

$$V_{HL} = \frac{V_{CM} Z_{HL}}{Z_{HL} + Z_{LE}} = \frac{V_{CM} \times 998}{-j \times 10\,000} = jV_{CM} \times 0.0998$$

Therefore

$$\text{CMRR (a.c.)} = \frac{V_{CM}}{V_{CM} \times 0.0998} = 10.02 = 20.02 \text{ dB}$$

Example 4.2

Here is the proof referred to on page 83. Figure 4.13(b) shows the equivalent circuit of a phase-sensitive detector when the FET is OFF and Figure 4.13(c) when it is ON. The r represents the ON resistance of an FET; the OFF resistance is sufficiently large to be ignored. It is required to prove that the small fraction of the input which reaches the noninverting input with F_1 ON can be compensated for by making the resistor to the noninverting input twice the other resistances and putting a similar FET, F_2 which is always ON, in the feedback.

Solution

When F_1 is OFF, both inputs and the output of the op-am are at almost exactly the same potential because the open loop gain is very large, so the gain is + 1. When F_1 is ON,

$$e_1 = \frac{(V_{out} - V_{in})}{R_1 + R_f + r} R_1 + V_{in}$$

$$= \frac{V_{out}R_1}{R_1 + R_f + r} + V_{in}\left(1 - \frac{R_1}{R_1 + R_f + r}\right)$$

$$e_2 = \frac{V_{in}\, r}{R_2 + r}$$

When the open loop gain is very large, $e_1 = e_2$:

$$\frac{V_{out}\, R_1}{R_1 + R_f + r} = V_{in}\left(\frac{r}{R_2 + r} + \frac{R_1}{R_1 + R_f + r} - 1\right)$$

which yields

$$\frac{V_{out}}{V_{in}} = \frac{r(R_1 + R_f + r)}{R_1(R_2 + r)} + 1 - \frac{R_1 + R_f + r}{R_1}$$

$$= \frac{r_1}{R_1}\frac{(R_1 + R_f - R_2)}{(R_2 + r)} - \frac{R_f}{R_1}$$

If we make $R_1 + R_f - R_2 = 0$, the expression becomes independent of r:

$$R_2 = R_1 + R_f$$

If $R_f = R_1$,

$$\frac{V_{out}}{V_{in}} = -1$$

Therefore, by making $\frac{1}{2}R_2 = R_f = R_1$, the gain will then be ±1, regardless of the ON resistance of F_1.

Example 4.3

A user required a portable accelerometer to measure low-frequency vibration. Before the detailed design could begin, it was necessary to decide whether a piezoelectric accelerometer could be used for the expected frequency and amplitude or whether it would be necessary to use a strain-gauge type. The preliminary calculations for the piezoelectric type were as follows.

An accelerometer with a sensitivity of 100 pC g^{-1} is used to measure an oscillation of frequency 2 Hz and amplitude 5 cm. Calculate the feedback capacitance, C_f, required in a charge amplifier to give an output voltage of 1 V peak to peak.

What is the minimum value of the feedback resistance R_f if the low-frequency response is 3 dB down at 0.25 Hz? With this resistance, how much amplifier input bias current can be tolerated to keep the offset less than 10 mV?

Comment on the practicability of this design. Take the acceleration due to gravity as 9.81 m s^{-2}.

Solution

Acceleration is related to displacement by the equation

$$a = -\ (2\pi f)x$$

Therefore,

peak–peak acceleration $a_{PP} = (2\pi 2)^2 \times 0.1$

$= 15.791$ m s^{-2}

$= 1.610$ g

charge output of accelerometer $Q_{PP} = 100 \times 1.610$

$= 161.0$ pC peak to peak

voltage output of amplifier $V = Q_{PP}/C_f = 1$ V

feedback capacitance $C_f = 161.0$ pF

From Section 4.4, the gain is 3 dB below the midfrequency value when $\omega R_f C_f = 1$. At 0.25 Hz, $R_f C_f = (2\pi \times 0.25)^{-1}$, therefore,

$$R_f = \frac{1}{2\pi \times 0.25 \times 161.0 \times 10^{-12}}$$

$$= 3955\ \text{M}\Omega$$

The offset voltage must be less than 10 mV, therefore the bias current must be less than

$$\frac{0.01}{3955 \times 10^{-6}} = 2.52\ \text{pA}$$

Note that this circuit could be made to work with a special high-impedance FET input op-amp and a carefully guarded R_f. The value of R_f is much greater than commonly available resistors, and might be reduced in practice by contamination with dirt or moisture. It would be much better to use a smaller R_f, larger C_f and an additional voltage amplification stage. Alternatively, use a strain-gauge accelerometer which has a frequency response right down to zero.

Example 4.4
A phase-locked loop comprises a voltage-controlled oscillator whose frequency is given by the equation $f = 400 + 20 \times V_C$, an exclusive-OR gate whose output levels are low = 1 V and high = 9 V, and a low-pass filter comprising 5 kΩ and 100 μF. The output of the filter is the control voltage V_C.

Calculate the lock range.

Solution
If $\phi = 0$, then the output of the detector equals the oscillator control voltage = 1 V, and the oscillator frequency = 400 + 10 × 1 = 410 Hz.

If $\phi = 180°$, then the output of the detector equals the oscillator control voltage = 9 V, and the oscillator frequency = 400 + 10 × 9 = 490 Hz.

This yields a lock range of 490 – 410 = 80 Hz.

CHAPTER 5

Sample and hold circuits

5.1 Introduction

A **sample and hold** circuit has several applications in data acquisition systems. It has both analogue and digital inputs and an analogue output. When the circuit is in the sample mode, the output follows the input and the circuit behaves like an op-amp, but when the digital (control) input puts the circuit into the hold mode, the output is held constant until the sample mode is resumed.

5.2 Applications

Typical applications are to

- hold the input signal to an A/D converter constant during conversion;
- permit effectively simultaneous sampling of several signals;
- hold a signal at the input of an A/D converter while the preceding multiplexer changes to a new channel;
- remove glitches from the output of a D/A converter (by using an open loop sample and hold which does not generate glitches of its own);
- demultiplex the output of a D/A converter.

5.3 Slew rate and aperture error

To understand why a sample and hold should sometimes precede an A/D, consider what the output of an A/D converter should represent. It should be either:

1. the mean value of the input over a specified interval; or
2. the value at a specified instant, and it is often required to measure a number of signals at the *same* instant.

(Note that the mean value over an unspecified or variable time is not a useful measurement.)

We can obtain the former by using an integrating converter, and the latter by using a converter that is so fast that the input does not change by more than one LSB during the conversion time.

Consider a sinusoidal input signal whose peak-to-peak amplitude $2A$ equals the full range of the converter, equivalent to 2^N bits:

$$v = A \sin(2\pi f_s t)$$

$$\frac{dv}{dt} = A 2\pi f_s \cos(2\pi f_s t)$$

$$\max \frac{dv}{dt} = A 2\pi f_s \tag{5.1}$$

Let us set the criterion that the signal should not change by more than 1 LSB during the conversion time, δt:

$$2\pi A f_s \, \delta t \leqslant \frac{2A}{2^N}$$

which gives

$$f_s \leqslant \frac{1}{2^N \pi \, \delta t} \tag{5.2}$$

If this condition is not met, then the change in voltage during δt is the **aperture uncertainty error**.

Figure 5.1 shows frequency plotted against the time taken to change by 1 LSB for various word lengths N. For example, consider the parameters of a common 12-bit converter with a conversion time of 25 μs. The maximum frequency which can be converted without any risk of aperture uncertainty is

$$(4096 \times \pi \times 25 \times 10^{-6})^{-1} = 3.108 \text{ Hz}$$

Admittedly, this is a worst-case error and not every conversion occurs at the maximum slew rate. Nevertheless, this frequency limit is so low as to be quite unacceptable. Taking another example, digital audio systems require 16-bit conversions at a sampling

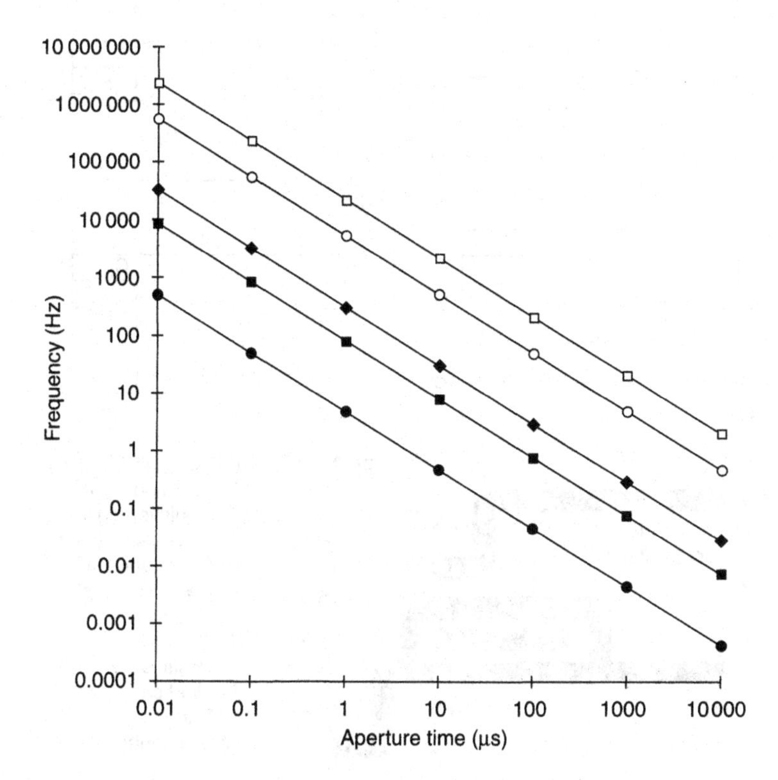

Figure 5.1 *Aperture time for less than 1 LSB change in signal: (●) N = 16; (■) N = 12; (◆) N = 10; (○) N = 8; (□) N = 4.*

rate of 40 kHz. This demands that the signal is sampled within 0.121 ns, although successive samples are 20 μs apart.

The problem is analogous to trying to photograph a moving object in a poor light. If the fastest available speed is not fast enough, the photographer will use a flash which illuminates the object for a shorter time than the shutter is open even at the highest shutter speed.

To determine the value of the input at the required instant, i.e. to sample it and hold it while digitization takes place, is the function of a sample and hold circuit. However, the alternative name track and hold gives a better description of its operation in this application.

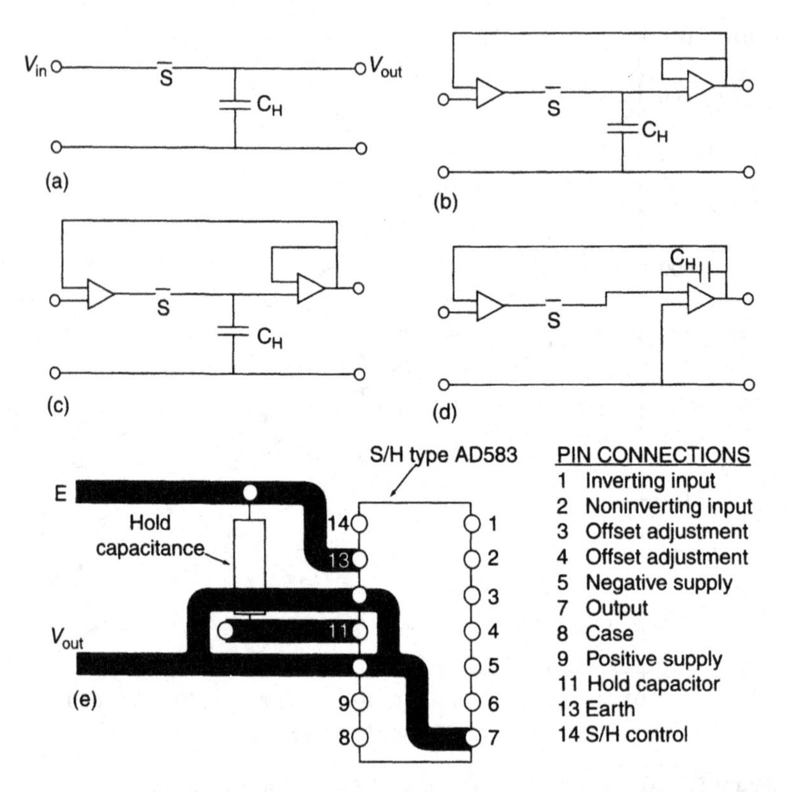

Figure 5.2 *Sample and hold circuits: (a) basic, impractical; (b) buffered, open loop; (c) buffered, closed loop; (d) virtual earth switch; (e) circuit board layout with guard ring.*

5.4 Basic design

At its simplest, a sample and hold circuit is as shown in Figure 5.2(a). The switch S is closed in the sample or track mode, and V_{out} follows V_{in}. In the hold mode the switch is open and V_{out} remains at the value of V_{in} at the instant the switch was opened. Practical implementations of this circuit will be described later.

5.5 Operation

Figure 5.3 illustrates the operation of the circuit:

- At A the control line changes to the command hold.
- At B the switch opens.

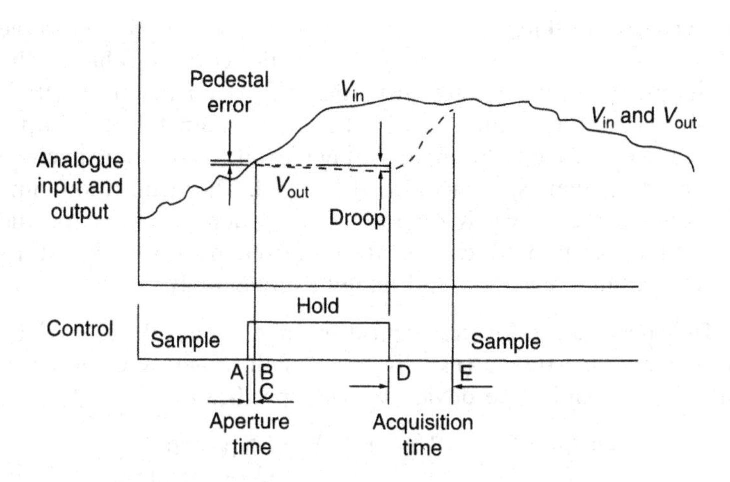

Figure 5.3 *Operation of a sample and hold.*

- At C the held signal appears at the output terminal.
- At D the control line reverts to sample.
- At E the output again follows the input within the specified limit.

5.6 Definitions

The following definitions are applicable to discussions of sample and hold circuits.

- **Sample mode** – Initially, the device is in the sample mode and the output voltage follows the input.
- **Aperture time** – This is from the hold command at A until the switch is fully open at B (Figure 5.3). When a sample and hold is in use, the aperture time replaces the conversion time in the equation for calculating aperture uncertainty error.
- **Settling time** – In some designs there is an additional delay from B to C before the acquired voltage appears at the output terminal. This is the settling time. The signal should not change by more than 1 LSB during the aperture time, but A/D conversion should not be started before the end of the settling time.
- **Hold mode** – From B/C to D the device is in the hold mode and, ideally, the output voltage of the sample and hold circuit will remain equal to the input at time B.

- **Acquisition time** – At D the control changes back to sample and the output begins to change to the current value of the input. It cannot do this instantaneously, and the time DE is the acquisition time, i.e. it is the time taken for the output voltage to follow the input voltage again to within a specified accuracy, that is, to acquire the signal. The acquisition time comprises the switch operating time, then the time for the output op-amp to charge stray capacitances initially at its maximum (slew) rate and finally exponentially.

The specification for the acquisition time defines the size of the step change in voltage (usually full or half range) and the accuracy. For example, one device has the specification

acquisition time to ±0.01% of 20 V = 1.5 μs typical
(3 μs maximum)

acquisition time to ±0.003% of 20 V = 2.5 μs typical
(5 μs maximum)

If the user has any choice of hold capacitor, the acquisition time will be quoted for specific values. The next sample cannot reliably be taken until the end of the acquisition time. Therefore, the acquisition time limits the **sampling rate** of the system.

5.7 Practical circuits

The basic circuit shown in Figure 5.2(a) would not work in practice. A complete sample and hold circuit includes level changers and switch drivers, and buffer amplifiers for both input and output. There are several popular configurations:

- **Open loop** – Input buffer, grounded capacitor and output buffer with open loop. This design is used for high-speed sample and holds (Figure 5.2(b)).
- **Closed loop** – Input buffer, grounded capacitor and output buffer with closed loop. When the switch is closed, the output tracks the input more accurately than the open loop configuration (Figure 5.2(c)).
- **Virtual earth** – Input buffer and output integrator with closed loop. The switch operates at earth potential, so leakage and droop are reduced (Figure 5.2(d)).

5.8 Errors

Offset and gain

If a sample and hold circuit is kept permanently in the sample mode, it behaves as an operational amplifier with nominal gain of ± 1 and is subject to offset and gain errors like any other op-amp.

Aperture jitter

This is the variation in the aperture time. In applications where several channels have sample and holds with a common command line, the sampling instants differ from one another by the **aperture jitter**, which is also called the **aperture uncertainty**, i.e. the uncertainty in the aperture time. Do not confuse this with the **aperture uncertainty error** which is the uncertainty in the output voltage caused by the existence of aperture time at all.

Droop

While the device is in the hold mode, the output may drift slightly at a rate which is called the **droop**. Droop may be in either direction and is caused by leakage of the switch or the hold capacitor, offset currents in the output buffer, or dielectric absorption. The effect of dielectric absorption is that the charge is initially held on the fast-moving charge carriers, but while the circuit is in the hold mode and the capacitor is isolated, the charge is shared between the fast and slowly moving charge carriers and the voltage is reduced. For this reason it is recommended to use polystyrene or polypropylene capacitors in sample and hold circuits. More details about dielectric absorption are given in Chapter 9, and in Gordon (1978).

In circuits, Figures 5.2(b) and (c), leakage of charge across the surface of the printed circuit board may be prevented by the track layout shown in Figure 5.2(e). The non-earthed end of the hold capacitor is surrounded by a ring of copper connected to the output terminal. Current will not leak from the capacitor to the ring because the output terminal holds the guard ring at the same potential. Any current which leaks to earth from the guard does not matter because it does not reduce the output voltage.

The droop rate increases rapidly with temperature.

Pedestal error

The change in output level between B and C is the **pedestal error**, also defined as **offset in sample mode – offset in hold mode**. It is due to charge from the hold capacitor being shared with the gate–drain capacitance of the switch. (The gate potential changes by several volts to turn the switch to off.) This loss of charge to a stray capacitance is also known as **charge suck-out**.

Feedthrough

Feedthrough is the component of output voltage which follows the input voltage when in hold mode. It is due to coupling through stray capacitances, particularly the source–drain capacitance of the switch.

5.9 Choice of hold capacitor

Although most sample and hold circuits are bought in integrated form, the user is left with the choice of hold capacitor. This must be of good quality and low dielectric absorption to avoid droop error, but the size of the capacitor is determined by a compromise between errors, as illustrated in Figure 5.4. Table 5.1 lists the

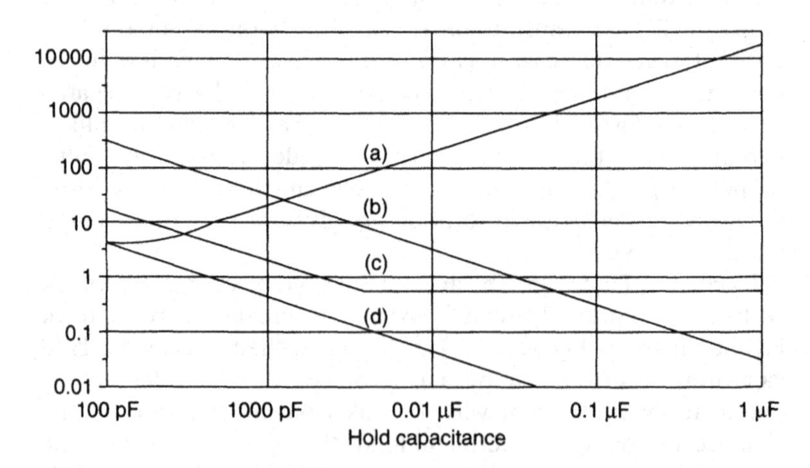

Figure 5.4 *Variation of parameters of a sample and hold with hold capacitance: (a) acquisition time 10 V step to 0.01% (μs); (b) droop rate (mV s^{-1}); (c) sample to hold offset (mV); (d) feedthrough, input 10 V (mV). Based on data sheet for AD582, used with permission.*

Table 5.1 *Dielectric absorption and insulation resistance*

Material	*Dielectric absorption (%)*	*Insulation resistance (MΩ)*
NPO ceramic	0.6	10^5
Stable ceramic	0.25	10^6
Mica	0.3–0.7	10^2
Polyester	0.5	10^4
Polycarbonate	0.35	10^5
Polypropylene	0.05	10^5
Polystyrene	0.05	10^6

values of dielectric absorption for commonly used capacitors, and Appendix C shows the other useful parameters.

To minimize acquisition time the hold capacitance should be small, but feedthrough, pedestal error and droop are all increased by a small capacitor. The manufacturer's specification will give typical values, from which the optimum for a particular application may be found. An example of this choice is given in Example 5.2.

5.10 Sampling converters

An A/D converter which has a built-in sample and hold circuit is catalogued as a **sampling converter**. These are becoming increasingly popular and relieve the circuit designer of the problem of matching the accuracy of the sample and hold to the accuracy of the converter.

5.11 Examples

Example 5.1

Calculate the maximum aperture time to digitize a 4 kHz signal to an accuracy of 1 LSB using a 10-bit converter with a conversion time of 10 μs.

If 16 samples are required in each period of the signal, what is the maximum acquisition time which can be used?

Solution
Given $f = (2^N \pi t_{AP})^{-1}$, where t_{AP} is the aperture time,

$$\frac{1}{4\,000} = 2^{10}\pi t_{AP}$$

$$t_{AP} = (4 \times 1024\pi)^{-1}\ \text{ms}$$

$$= 77.7\ \text{ns}$$

Sampling interval t_{SAM} for 16 samples in each period:

$$t_{SAM} = (16 \times 4000)^{-1}$$

$$= 15.625\ \mu\text{s}$$

As 10 μs are required for conversion, and 0.077 μs are required for aperture time,

$$15.625 - 10 - 0.077 = 5.548\ \mu\text{s}$$

are available for acquisition.

Therefore, allow up to 5 μs in the sample and hold specification.

Example 5.2
Many sample and hold ICs use an external hold capacitor, and the choice of the hold capacitor gives the user control over some of the parameters.

The specification of one type of sample and hold includes the following, where C_H is the capacitance of hold capacitor in pF:

aperture time = 0.05 μs

acquisition time to ½ LSB of 12-bit converter = $6 + 0.022C_H$ μs

droop current = 100 pA

sample to hold offset (pedestal error) = $0.0005 + (5/C_H)$ V

This sample and hold is to be used with a 16-channel multiplexer having a settling time of 2 μs and a 12-bit, 10 V A/D having a conversion time of 25 μs. The sampling rate is 1000 samples per second. All channels are to be sampled simultaneously.

Calculate the maximum and minimum hold capacitance so that neither acquisition, droop nor pedestal error exceeds ½ LSB.

What dielectric material would you use in this capacitor?

Solution

The time required to measure 16 channels is the hold time:

$$\begin{aligned} t_H &= t_{AP} + 16(t_{CONV} + t_{MUX}) \\ &= 0.05 + 16(25 + 2) \\ &= 432.05\ \mu s \end{aligned}$$

where t_{AP} and t_{CONV} are the aperture time and A/D conversion time, respectively.

The time between samples is 1000 μs; therefore, the time available for acquisition is

$$1000 - 432.05 = 567.95\ \mu s$$

The upper limit to the size of the hold capacitor is given by the acquisition time and the lower limit by the droop rate or the sample-to-hold offset. In order to acquire the signal to ½ LSB as specified within the time available:

$$t_{AQ} = 6 + 0.022C_H \leqslant 567.95$$

$$C_H \leqslant 25\,543\ \text{pF}$$

$$\frac{1}{2}\,\text{LSB} = \frac{V_{FS}}{2 \times 2^N}$$

where V_{FS} is the full-scale voltage and N is the number of bits.

Let the droop current be i_{DR} amps; then the droop rate is

$$\frac{i_{DR}}{C_H}\ \text{V s}^{-1}$$

The fall in voltage during hold time is

$$\frac{i_{DR}\, t_H}{C_H}\ \text{V}$$

Therefore, the lower limiting value of C_H is given by

$$\frac{V_{FS}}{2 \times 2^N} \geqslant \frac{i_{DR}\, t_H}{C_H}$$

$$C_H \geqslant \frac{i_{DR}\, t_H}{V_{FS}}\, 2^{N+1}$$

$$\geqslant 10^{-10}\,\frac{432 \times 10^{-6}}{10} \times 8192\ \text{F}$$

$$\geqslant 432 \times 10^{5} \times 8192\ \text{pF}$$

$$\geqslant 35.4\ \text{pF}$$

The sample to hold offset is $0.0005 + (5/C_H)$ V.

Also, the limiting value of C_H is given by

$$\frac{V_{FS}}{2 \times 2^N} \geqslant 0.0005 + \frac{5}{C_H}$$

$$\frac{10}{8192} - 0.0005 \geqslant \frac{5}{C_H}$$

$$0.000\,720\,7 \geqslant \frac{5}{C_H}$$

$$C_H \geqslant 6938\ \text{pF}$$

Thus the lower limit is set by the offset, not the droop, and C_H must lie between 6938 and 25 543 pF; 10 000 pF, i.e. 10 nF, would be suitable.

A low absorption dielectric such as polypropylene should be used.

Example 5.3

This example considers the use of a sample and hold in a system. Systems are discussed in more detail in Chapter 10.

Figure 5.5 shows a 10-bit data acquisition system, in which V_1 and V_2 are input signals of frequency 3 kHz with any relative phase and any amplitude up to 10 V. Calculate:

- maximum quantization error;
- maximum uncertainty due to signal changing during digitization;
- maximum skew error.

Show where a sample and hold circuit or circuits could be added to the system. What are the maximum aperture time, acquisition time and droop rate such that the sample and hold errors are no larger than the maximum quantization error. (Assume that settling

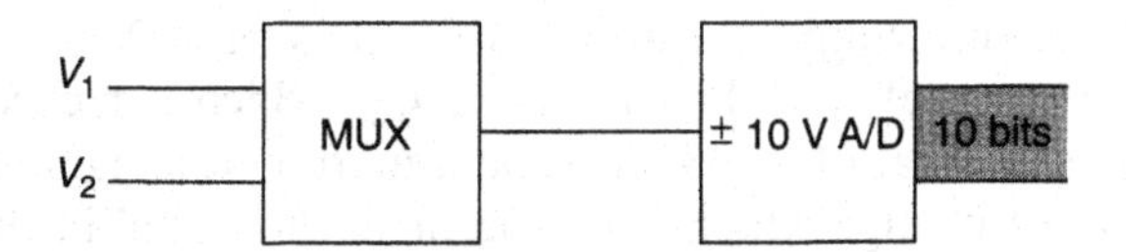

Figure 5.5 *Example 5.3.*

time = aperture time = 1 μs, and that analogue to digital conversion time = 15 μs.)

Solution

The maximum amplitude is ± 10 V, so the range of the converter is 20 V. This is represented by 10 bits, but as the highest usable code is all ones, the range is only ± 511; so one LSB = 20/1022 = 19.5695 mV.

Quantization error is defined on page 213. Maximum quantization error is

$$\tfrac{1}{2}\,\text{LSB} = 9.7847$$

$$= 9.78\ \text{mV (to three significant figures)}$$

From (5.1) the maximum rate of change of signal is

$$2\pi fA = 2\pi 3000 \times 10\ \text{V s}^{-1}$$

$$= 188\,495\ \text{V s}^{-1}$$

Without a sample and hold circuit

Maximum change during digitization is

$$188\,495 \times 15 \times 10^{-6} = 2.827\,43$$

Allowing 1 μs for settling time and 15 μs for digitization, the minimum time between successive samples is 16 μs. The worst-case skew error occurs if the signals are sampled when one is changing at the maximum rate. The skew error is then

$$188\,495 \times 16 \times 10^{-6} = 3.015\,92$$

$$= 3.02\ \text{V (to three significant figures)}$$

Not many pairs of samples will have skew errors as large as this.

If the sampling rate is slower and the samples from both channels are evenly spaced, the skew error could be much worse.

The minimum sampling rate to avoid aliasing is $2f_{sig} = 6$ kHz.

For two channels the throughput is $4f_{sig}$, which corresponds to a sampling interval of 83.33 μs or one-quarter of a signal period.

The largest change which could occur in the signal in this time is from 0.707 V_{max} to −0.707 V_{max} or vice versa, i.e. 1.414 V_{max} which, in this example, is 7.07 V.

With a sample and hold circuit before the A/D converter

The aperture error uncertainty would be reduced from 2.827 43 to 188 495t_{AP}. For this to be less than the quantization error, 188 495$t_{AP} \leq 0.009\,784\,7$; $t_{AP} \leq 51.9$ ns.

The skew error would be unchanged; to reduce it, the sample and hold must be placed before the multiplexer.

With a sample and hold circuit before the multiplexer

Both channels would be held at the same moment and the aperture uncertainty error would be reduced as above. The skew error would be reduced to

$$188\,495(t_{APA} - t_{APB})$$

The difference in aperture times of the two sample and hold circuits $(t_{APA} - t_{APB})$ is known as the aperture uncertainty or aperture jitter. It is always less, usually much less, than the aperture time, so the skew error would be less than the aperture error.

With a sample and hold circuit before the multiplexer, the sequence of operations is hold, convert, multiplex, convert (multiplex and sample).

The time required is

$$0.1 + 2 \times 15 + 1 + t_{AQ} = 31.1 + t_{AQ}$$

(This assumes that the acquisition time is more than the multiplexer settling time of 1 μs.)

If the sampling rate of each channel is 6 kHz, the time available for acquisition is

$$\frac{10^6}{6000} - 31.1 = 135.6 \text{ μs}$$

If the sampling rate is higher, as it probably will be, the time available for acquisition is correspondingly less.

The sample and hold circuit must hold the signal constant to within ½ LSB for 31.1 μs, so the maximum droop rate which can be tolerated is

$$\frac{9.7847 \times 10^{-3}}{31.1 \times 10^{-6}} = 314.62$$

$= 314\ \mathrm{V\ s^{-1}}$ (to three significant figures)

CHAPTER 6

Multiplexers

6.1 Introduction

A **multiplexer, selector** or **scanner** is essentially a set of switches used to connect one of a number of alternative inputs to a single device in sequence, in order to make more efficient use of an expensive component. A digital multiplexer may be used to select one digital word or byte from a number of sources, but here we are primarily concerned with analogue multiplexers as these are much more common in data acquisition systems. In these systems, the most expensive component may be the signal conditioning, but it is most likely to be the high resolution A/D converter and the computer. Therefore, it is very common to multiplex a number of sensors to one converter. How this affects the positions of sample and hold circuits and the timing of the various operations is discussed in Chapter 10, but this chapter considers a multiplexer as a component by itself. A system may also include a **demultiplexer** to connect the output of a D/A converter to a number of different points.

In manually operated instruments the multiplexing operation can be done by one or more multiposition switches, but in an automated system the switches must be operated by a digital signal and so a number of separate switches are employed with their outputs connected together as shown in Figure 6.1.

In this configuration it is important that the switches break-before-make, i.e. each switch must open before the next one closes so that the signal sources, which are low impedance, are never connected together.

Multiplexer switches may be either reed relays or FET switches. A comprehensive account is given in the *Switching Handbook* (Keithley, 1995).

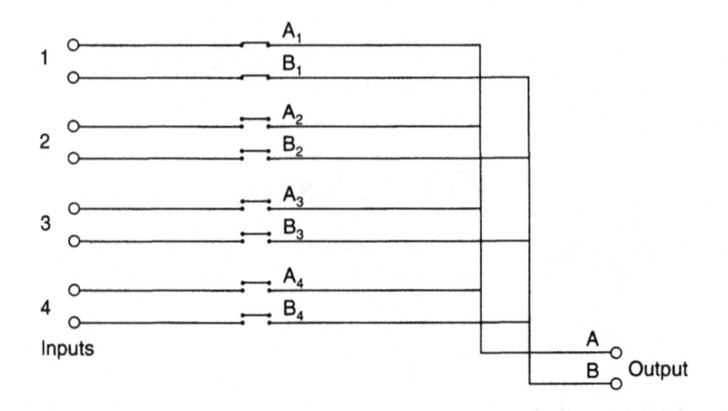

Figure 6.1 *Multiplexer circuit.*

6.2 Number of switches required

If the signals are single ended and share a common ground line, then one switch for each channel is, theoretically, sufficient.

Even if the signals are not differential, better common-mode rejection can be achieved by using two switches in each channel and a differential input to the next stage. Integrated circuit switches are often specified as, for example, '16 channels single ended or 8 channels differential', and if using a differential multiplexer it obviously makes sense to have both switches on the same chip and therefore at the same temperature.

For the highest accuracy, the measuring instrument will have a guard box which should be connected as closely as possible to the low-potential side of the signal source. In this case, the guard connections also need to be switched.

6.3 Reed relays

A reed switch is illustrated in Figure 6.2. It consists of two or three thin, flat strips of ferrous metal, the so-called reeds, sealed into a glass tube which is filled with inert gas. The overlapping ends of the reeds carry contacts made of a noble metal, such as gold, silver, rhodium or tungsten. When a magnetic field is applied to the switch parallel to the reeds, they become magnetized and the overlapping ends attract each other and operate the switch. A **reed relay** comprises a reed switch enclosed in an operating coil.

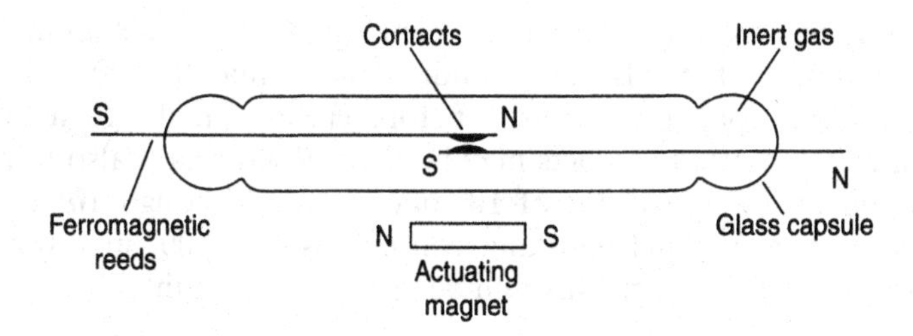

Figure 6.2 *Reed switch.*

Reed relay coils can be obtained for operation on voltages between 5 and 50 V, and the coil power consumption ranges from tens of milliwatts to almost one watt. A particularly useful type is the dual in line (DIL) relay, which looks like an integrated circuit with thicker than usual plastic.

Normally open contacts are known as form A, normally closed are form B, and changeover are form C.

A diode must be connected across the relay coil in the direction which is high resistance to the coil supply. When the supply is disconnected, the energy stored in the inductance of the coil will discharge through the diode. Protective diodes are usually incorporated in DIL relays, so the coils must be connected to the supply in the correct polarity. Larger relays may require the user to provide the diode and may store sufficient energy to burn out the coil if it is omitted, so it pays to check that the diode is in circuit.

6.3.1 Advantages of reed relays

The ON resistance of a reed switch is of the order of 0.1 Ω, and the OFF resistance is of the order of 10^{10} Ω, giving an OFF:ON ratio of 10^{11}, which is several orders greater than can be achieved with solid-state switches. This makes them particularly suitable for multiplexing low-level signals such as are obtained from strain gauges or thermocouples. Furthermore, the contacts are completely isolated from the control circuit and a potential of several hundred volts may be applied between the coil and contacts.

6.3.2 Disadvantages of reed relays

Although reed relays are fast as relays go, they are very much slower than solid-state switches. The operation time is typically a few milliseconds, thus limiting the scanning rate to a maximum of a few hundred channels per second. Contacts may 'bounce', that is close then open momentarily before closing finally. To avoid this, mercury-wetted contacts may be used. Reed relays also have a limited life of the order of 10^9 operations. Although 10^9 is a large number, a switch working continuously at 100 operations per second would reach this number in only 4 months!

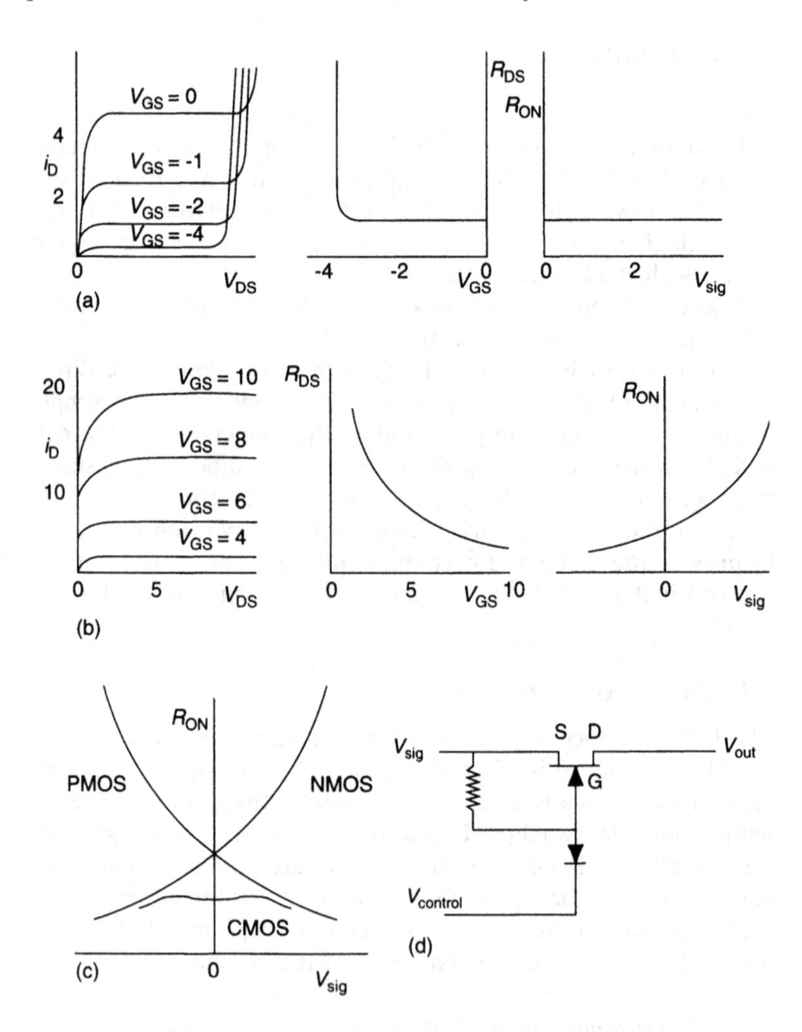

Figure 6.3 *FET switches: (a) n-channel JFET characteristics; (b) n-channel MOSFET characteristics; (c) CMOSFET characteristics; (d) control circuit for n-channel JFET.*

Thermoelectric emfs may be generated at junctions in either solid-state or reed switches, but the heat generated by the coil may increase the effect in reeds. Horizontal mounting will help to keep both ends at the same temperature, although mercury-wetted types may have to be mounted vertically.

6.4 FET switches

Field effect transistors can be used either as amplifiers or as switches. They are more suitable for multiplexer switches than bipolar transistors because they do not introduce an offset. There is almost no voltage drop across an FET when the current is zero. A detailed review of solid-state switches is given by Bolger (1980a,b) and many of the devices described are still available.

6.4.1 Junction FETs (JFETs)

The current flowing through the channel from the source to the drain is controlled by the potential of the gate which is made of the opposite polarity semiconductor to the channel. In most FETs, current can flow between the source and drain in either direction. The general form of the characteristics of an *n*-channel, depletion-mode JFET are shown in Figure 6.3(a).

Note that when the gate–source voltage V_{GS} is zero, the drain current is high and the drain–source resistance R_{DS} is low. The JFET is then said to be ON and R_{DS} is designated R_{ON}. Beware that all the switches in a JFET multiplexer are ON until power is applied, and the signal inputs might be connected to each other if they were connected to the multiplexer before the power. The ON resistance of a JFET is independent of the signal voltage and is of the order of tens of ohms for *n*-channel and a few hundred ohms for *p*-channel types.

Normal logic levels are too small and the wrong polarity to control an *n*-channel FET, so a driver transistor must be used. A driver packaged with an FET is termed a switch. A control circuit for an *n*-channel JFET is shown in Figure 6.3(d). When the control voltage is zero or positive, the gate is tied to the source through R and the FET is ON. When the control voltage is negative, the diode conducts, the gate is held negative and the FET is OFF. The diode prevents the control voltage making the gate positive to the source.

6.4.2 MOSFETs

A metal oxide semiconductor field effect transistor or MOSFET has a gate made of metal which is insulated from the channel by an extremely thin layer of metal oxide. In the depletion mode, the voltage on the gate V_{GS} reduces the number of current carriers in the channel and the characteristics are very similar to those of a JFET. Conversely, an *n*-channel enhancement mode FET is ON when the gate is held at a constant positive potential (or negative for *p*-channel). The gate voltage V_{GS} does not follow the source voltage as it does in the JFET circuit. As the signal voltage V_{sig} goes more positive, V_{GS} goes less positive and increases R_{DS}. The ON resistance thus varies with V_{GS} and therefore with the analogue signal. This limits the amplitude of the signal the switch can handle. The characteristics of an *n*-channel enhancement MOSFET are illustrated in Figure 6.3(b) together with the variation of R_{ON} with V_{sig}.

6.4.3 CMOSFETs

To reduce the variation is ON resistance, a *p*-channel and an *n*-channel MOSFET are connected in parallel making a complementary pair. It is, of course, necessary to provide complementary control signals as well. These are all included in a CMOS switch. CMOS switches can be used for signals of either polarity, and Figure 6.3(c) shows the variation of R_{ON} with signal for both *n*- and *p*-channel MOSFETs and for a CMOS switch.

6.4.4 Equivalent circuit of a JFET

A JFET may be represented by the equivalent circuit shown in Figure 6.4. The components r_{GS} and r_{GD} are reverse-biased diode junctions, and C_{GS}, C_{GD} and C_{SD} are the stray capacitances from gate to source, gate to drain and source to drain, respectively. G_{GS} and C_{GD} can inject charge into the signal circuit (source to drain) as the control voltage changes and, because the control voltage may be many times larger than the signal voltage, this can cause problems.

Data sheets do not quote self-capacitances directly but as input and output strays. The input stray capacitance is measured from gate to source with C_{SD} shorted by a large capacitor:

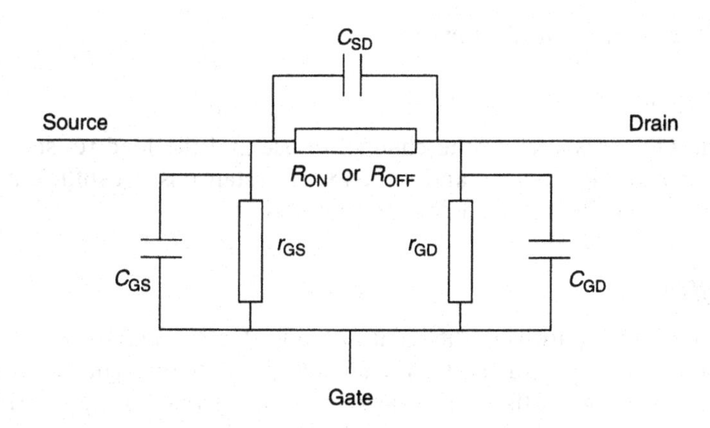

Figure 6.4 *Equivalent circuit of a JFET.*

$$C_{ISS} = C_{GSS} = C_{GS} + C_{GD} \tag{6.1}$$

The output stray capacitance is measured from gate to drain with C_{GS} shorted by a large capacitor:

$$C_{OSS} = C_{RSS} \approx C_{GD} \tag{6.2}$$

The values of these capacitances vary with voltage, but they are of the order of a few picofarads. There is a trade-off between the self-capacitances and R_{ON} in that FETs with a low R_{ON} tend to have larger self-capacitances. For example, in type 2N4393

$$C_{ISS} = 8 \text{ pF}, C_{RSS} = 4 \text{ pF and } R_{ON} = 100\ \Omega$$

and in type 2N4861

$$C_{ISS} = 18 \text{ pF}, C_{RSS} = 8 \text{ pF and } R_{ON} = 60\ \Omega$$

Both these devices are *n*-channel JFETs and maximum values are quoted. In practice, other stray capacitances on the circuit board may increase the effective value of any of these self-capacitances.

6.5 Errors in multiplexers

Attenuation

The ON resistance of the channel in use and the load resistance form a voltage divider, and if the load resistance is not sufficiently large, a buffer amplifier will be required.

Offset

If V_G and V_D are very different in the OFF state, leakage current through the R_{GD} of the OFF channels can flow through the load and cause an 'offset'. Leakage currents increase rapidly with temperature.

Glitches

As a switch is turned ON or OFF, the step function change of V_G can transmit a spike through C_{GD} and cause a glitch on the output.

Settling time

The settling time is made up of the time to turn the current on and the time to charge stray capacitances, including the input capacitance of the load, and is in the range 50 ns to 3 μs.

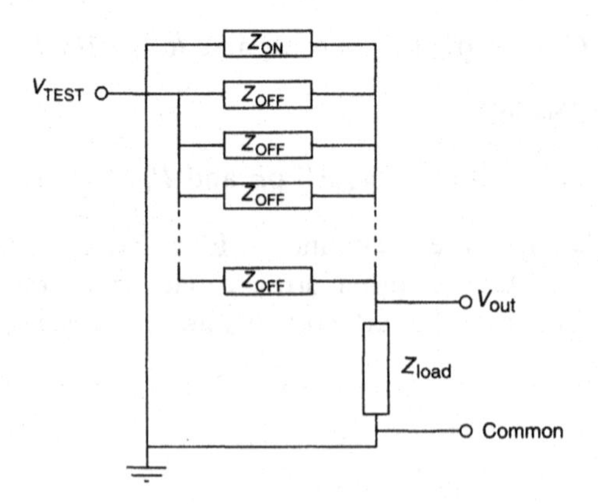

Figure 6.5 *Crosstalk.*

Crosstalk

Crosstalk is the leakage of signals between channels, and is most easily understood by considering the test circuit used to measure **crosstalk attenuation** shown in Figure 6.5. One input channel only is switched ON and earthed; all the other channels are switched OFF, but connected to the test voltage. If the multiplexer were perfect, there would be no voltage at the output, so whatever voltage is measured at the output has come through the OFF channels and is an error. The crosstalk attenuation is the test voltage divided by the resulting error and is expressed in decibels.

The test voltage drives current through all the OFF channels to earth through the load and also back through the ON channel:

$$Z_{ON} = R_{source} + R_{ON} \parallel C_{SD}$$

where $\parallel$ means 'in parallel with',

$$Z_{OFF} = R_{source} + R_{OFF} \parallel C_{SD}$$

$$Z_{load} = R_{load} \parallel C_{load} + NC_{GD}$$

Impedance through the OFF channels to the V_{out} terminal = $Z_{OFF}/(N-1)$.

Impedance from the V_{out} terminal to earth = Z_{ON} in parallel with Z_{load}:

$$Z_{ON} \parallel Z_{load} = \frac{Z_{ON}\, Z_{load}}{(Z_{ON} + Z_{load})}$$

Therefore,

$$\frac{V_{out}}{V_{TEST}} = \frac{Z_{ON} \parallel Z_{load}}{Z_{ON} \parallel Z_{load} + Z_{OFF}/(N-1)}$$

$$= \frac{(N-1)\,(Z_{ON}\, Z_{load})}{(N-1)\,(Z_{ON}\, Z_{load}) + (Z_{ON} + Z_{load})\, Z_{OFF}} \qquad (6.3)$$

Normally, Z_{ON} is much less than Z_{load} and this simplifies to

$$\frac{(N-1)\, Z_{ON}}{(N-1)\, Z_{ON} + Z_{OFF}} \qquad (6.4)$$

The crosstalk attenuation is given by

$$\text{crosstalk attenuation} = 20 \log_{10} \frac{V_{\text{TEST}}}{V_{\text{out}}}$$

$$= 20 \log_{10} \left[1 + \frac{Z_{\text{OFF}}}{Z_{\text{ON}}(N-1)} \right] \quad (6.5)$$

therefore

$$\text{crosstalk attenuation} \approx 20 \log_{10} \left[\frac{Z_{\text{OFF}}}{Z_{\text{ON}}\,(N-1)} \right] \text{dB} \quad (6.6)$$

At high frequencies C_{SD} affects the crosstalk by reducing Z_{OFF}.
A detailed description of crosstalk is given by Givens (1979).

Two-stage multiplexing

If a large number of channels are to be multiplexed to the same converter, crosstalk may be reduced by multiplexing in two stages (or tiers). For example, 80 channels may be reduced to eight, by eight 10-channel multiplexers, and then to 1 by a further 10-channel multiplexer.

By a similar argument, it can be shown that for a multiplexer having M channels in the first tier and N channels in the second,

$$\text{crosstalk attenuation} = 20 \log_{10} \left[\frac{Z_{\text{OFF}}}{Z_{\text{ON}}\,(M+N-2)} \right] \quad (6.7)$$

6.6 Examples

Example 6.1
Calculate the crosstalk attenuation at d.c. of a 16-channel multiplexer made from switches with an ON resistance of 100 Ω and an OFF resistance of 10^7 Ω (Figure 6.6). The input resistance of the next stage is 1 MΩ.

How could the crosstalk attenuation be improved?

Solution
From (6.6)

$$\text{crosstalk attenuation} = 20 \log \left[\frac{Z_{\text{OFF}}}{Z_{\text{ON}}\,(N-1)} \right] \text{dB}$$

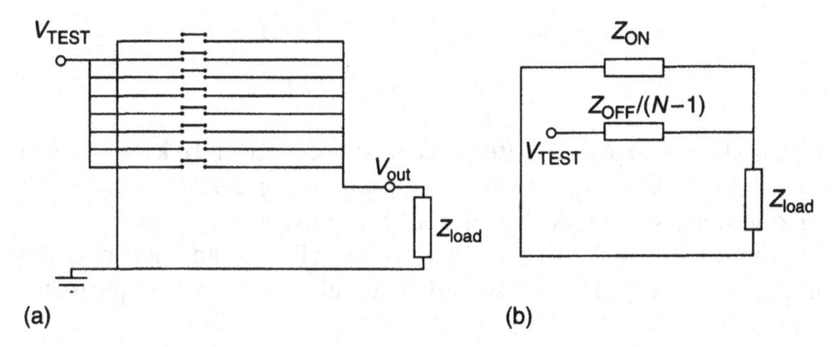

Figure 6.6 *Example 6.1.*

where N is the number of channels, Z_{OFF} is the impedance of the switch in the OFF state, and Z_{ON} is the impedance of the switch in the ON state. Substituting values gives

$$\text{crosstalk attenuation} = 20 \log \left(\frac{10^7}{100 \times 15} \right)$$

$$= 76.5 \text{ dB}$$

The crosstalk attenuation could be increased by using two stages of four channels each. Using (6.7) the crosstalk attenuation would then become

$$\text{crosstalk attenuation} = 20 \log_{10} \left[\frac{(Z_{OFF})}{Z_{ON}(M + N - 2)} \right]$$

$$= 20 \log \left(\frac{10^7}{100 \times 6} \right)$$

$$= 84.4 \text{ dB}$$

Example 6.2

An eight-channel multiplexer is made of CMOS switches to the following specification:

ON state: $R_{DS} = 75\ \Omega$

$C_{DS} = 1$ pF

OFF state: $R_{DS} = 4 \times 10^9\ \Omega$

$C_{DS} = 1$ pF

Load: $R_L = 4\ \text{M}\Omega$

$C_L = 50\ \text{pF}$

Calculate the crosstalk attenuation at d.c. and at 10 kHz. With a full-range 10 kHz signal, which would be the greater: the crosstalk or one least significant bit of a 12-bit converter?

Would you consider these switches suitable for an eight-channel multiplexer at 10 kHz or would reed relays give better performance?

Solution

From (6.6)

$$\text{crosstalk attenuation} = 20 \log \left[\frac{Z_{OFF}}{Z_{ON}(N-1)} \right] \text{dB}$$

where N is the number of channels, Z_{OFF} is the impedance of the switch in the OFF state, and Z_{ON} is the impedance of the switch in the ON state.

At d.c.,

$$\text{crosstalk attenuation} = 20 \log \left(\frac{4 \times 10^9}{75 \times 7} \right) = 7.619 \times 10^6$$

$$= 137.6\ \text{dB}$$

At 10 kHz the drain–source reactance of each switch X_{DS} is

$$X_{DS} = \frac{1}{2\pi 10^4 C_{DS}} = \frac{10^8}{2\pi}$$

$$= 15.92\ \text{M}\Omega$$

Now X_{DS} is much less than R_{DSOFF}, so $Z_{OFF} = 15.9\ \text{M}\Omega$.

The load reactance X_L is

$$X_L = \frac{1}{2\pi 10^4 C_L} = \frac{1}{2\pi 10^4 \times 50 \times 10^{-12}}$$

$$= 0.3183\ \text{M}\Omega$$

so Z_L is still much greater than Z_{ON}. Therefore,

$$\frac{V_{TEST}}{V_{OUT}} = \frac{15.9 \times 10^6}{75 \times 7} = 30\,286$$

and

$$\text{crosstalk attenuation} = 89.6 \text{ dB}$$

The root mean square (rms) of a full-range signal is $V_{FS}/2\sqrt{2}$. From a full-range signal at 10 kHz,

$$\text{crosstalk} = \frac{V_{FS}}{2\sqrt{2} \times 30\,286} = \frac{V_{FS}}{85\,662}$$

and one least significant bit is

$$\text{LSB} = \frac{V_{FS}}{4096}$$

Therefore, the crosstalk is less than 1 LSB and this switch is satisfactory for this application.

A 10 kHz signal would have to be sampled at 20 kHz minimum. If we assume that the multiplexer changes channels after each sample, the switching rate will be 8 × 20 000 operations per second which is much too fast for reed switches.

Example 6.3

A manufacturer of data loggers offers two types of 10-channel multiplexer card to the following specifications:

Switch	*Offset*	*Settling time*	*ON channel resistance*	*OFF channel resistance*
Reed	± 10 μV	2 ms	100 mΩ	10^8 Ω
FET	± 1 mV	20 μs	5000 Ω	10^6 Ω

Calculate the crosstalk attenuation for each type.

One hundred identical strain-gauge bridges are connected to an analogue to digital converter by a two-tier multiplexer, one tier of reed switches and one of FETs. The maximum signal from each strain-gauge bridge is 10 mV. The A/D converter has a full range input of 2 V, a conversion time of 25 μs and a resolution of 10 bits.

Sketch the configuration of the system showing where you would use amplifiers and what their gain would be. Consider the switch offsets as a random error which cannot be compensated.

Calculate the minimum time needed to sample all channels and the maximum slew rate of the input signals such that the slew error is less than ½ LSB.

Discuss how the circuit could be modified to improve the sampling rate.

Solution

The equation for crosstalk attenuation is derived on page 119. From (6.6)

$$\text{crosstalk attenuation} = 20 \log \left[\frac{Z_{OFF}}{Z_{ON}\,(N-1)} \right] \text{dB}$$

where N is the number of channels, Z_{OFF} is the impedance of the switch in the OFF state, and Z_{ON} is the impedance of the switch in the ON state.

For reed cards,

$$\text{crosstalk attenuation} = 20 \log \left(\frac{10^8}{9 \times 0.1} \right) = 161 \text{ dB}$$

For FET cards,

$$\text{crosstalk attenuation} = 20 \log \left(\frac{10^6}{9 \times 5000} \right) = 27 \text{ dB}$$

The configuration of the system is given by Figure 6.7. The low offset of reed switches makes them suitable for use before amplification.

$$\text{gain required} = \frac{2}{10 \times 10^{-3}} = 200$$

Timing (in microseconds):

A/D conversion time	25
FET switch settling	20
total for one channel	45
total for 10 channels	450
reed switch settling	2 000

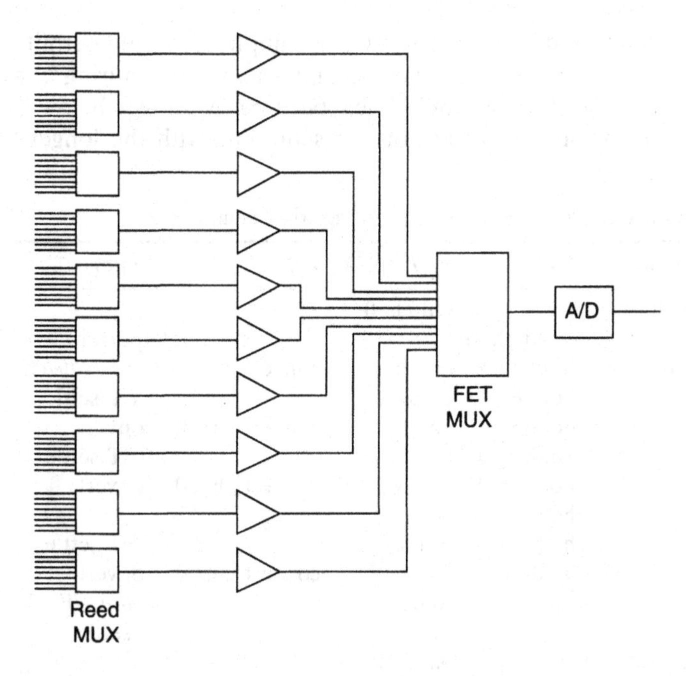

Figure 6.7 *Example 6.3.*

total for 10 channels	2 450
total for 100 channels	24 500 = 24.5 ms

The slew rate is calculated as follows: range = 2 V; number of bits = 10; therefore,

$$\mathrm{LSB} = \frac{2}{2^{10}} = 1.953\ \mathrm{mV}$$

Maximum slew rate for slew error $\leq$ ½ LSB is

$$\frac{1.953}{2 \times 0.0245} = 39.9\ \mathrm{mV\ s^{-1}}$$

Two alternatives to improve the sampling rate are as follows:

- If both tiers of the multiplexer were solid state, the switching would be much faster, but it would be necessary to use amplifiers before both tiers because of the offset in the FET switches, which is of the order of 1 LSB. Thus 100 amplifiers would be needed instead of 10.

- Sample and hold circuits after the amplifiers would permit each reed to move to the next channel and settle during conversion. The timing would then become as shown below (concurrent operations are on the same line with the longer time in italics).

Data acquisition sequence and timing for Example 6.3

Reed MUX	*S/H*	*FET MUX*	*A/D*	*Times*
On ch. 0		On ch. 0		
	Hold ch. 0			aperture
to ch. 1	holding 0	on ch. 0	convert ch. 00	*reed settle*/convert
	holding 0	to ch. 1		FET settle
	holding 0	on ch. 1	convert ch. 10	convert
	holding 0	to ch. 2		FET settle
	holding 0	on ch. 2	convert ch. 20	convert
	holding 0	etc.		
	holding 0	to ch. 9		FET settle
	holding 0	on ch. 9	convert ch. 90	convert
	sample	to ch. 0		*acquire*/FET settle
On ch. 1		on ch. 0		
	Hold ch. 1			aperture
to ch. 2	holding 1	on ch. 0	convert ch. 01	*reed settle*/convert
	holding 1	to ch. 1		FET settle
	holding 1	on ch. 1	convert ch. 11	convert

and so on until all MUXs are on ch. 9 and ch. 99 has been converted

Total times: $10(t_{AP} + t_{AQ} + t_{RS}) + 100(t_{CON} + t_{FS})\mu s$

$10(t_{AP} + t_{AQ} + 2000) + 100(25 + 20)\mu s$

$> 20 \text{ ms} + 4.5 \text{ ms}$

The 10 settling times for the reed switches take much longer than the 100 FET settling times and conversions, so the total time to digitize all the channels is more than 20 ms which is not a great improvement on the 24.5 ms required by the simpler circuit in Figure 6.6.

CHAPTER 7

Elements of analogue to digital and digital to analogue converters

7.1 Introduction

Measurement is essentially a process of comparing one quantity (the measurand) with another (the standard) and expressing the result as a ratio. It follows that the accuracy of the measurement cannot be greater than the accuracy to which the standard is known. All A/D and D/A converters represent an analogue voltage as the product of a reference voltage and a pure number, and the number is expressed in a binary code. This chapter discusses voltage reference devices and the representation of a number in a binary code as a necessary precursor to the separate chapters on A/Ds and D/As which follow. Details of the theory of measurement were given in Chapter 2.

7.2 Voltage references

The term **voltage reference** is used for the device which provides the **reference voltage**. The earliest A/Ds used standard cells, but all modern converters use electronic devices, either Zener diodes or bandgap references.

7.2.1 Zener diodes

If the electric field in a reverse-biased diode exceeds 2×10^7 V m^{-1}, it may remove electrons from covalent bonds, which is the mechanism of **Zener breakdown**. In heavily doped diodes this occurs at voltages greater than 6 and the breakdown voltage has a negative temperature coefficient. A similar effect occurs in less heavily doped diodes except that the current is increased by **avalanche**

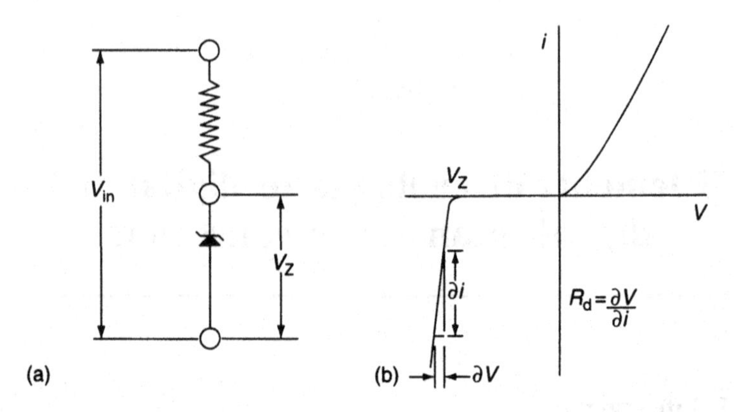

Figure 7.1 *Zener diode: (a) circuit; (b) characteristics.*

breakdown. This has a positive temperature coefficient of the order of 0.1% K^{-1}.

The name **Zener diode** is used indiscriminately for devices with current–voltage characteristics of the form in Figure 7.1, whatever the mechanism of the breakdown. The slope of the part of the curve beyond the breakdown voltage is the **dynamic resistance**:

$$R_d = \frac{\partial V}{\partial i} \tag{7.1}$$

The temperature coefficient may be compensated by combining a positive coefficient Zener with a forward-biased diode (which has a negative coefficient). This gives an S-shaped characteristic and the fine print of the data sheet must be consulted to determine the coefficient in a particular circuit.

Some voltage references include a transistor acting as a heater and keeping the substrate of the Zener at a thermostatically controlled temperature above ambient. These are said to be temperature stabilized.

A **buried Zener** has the junction buried below the surface so that it will not be affected by strain in the leads, or surface contamination.

Zener references become more stable with age. Datron (now Wavetek–Datron), for example, aged their Zeners for 2 years before building them into digital voltmeters. They also used eight Zeners in series to statistically reduce random voltage fluctuations (Prophet, 1983; Datron, 1987).

Examples of low temperature coefficient devices are:

IN 827 6.2 V: (temperature compensated) temperature coefficient $= 0.001\%\ °C^{-1} = 10$ ppm (parts per million)

IN 8241 6.2 V: temperature coefficient = 5 ppm $°C^{-1}$

LM199 6.95 V: temperature coefficient = 0.5 ppm $°C^{-1}$; temperature stabilized

LTZ1000 7.2 V: temperature coefficient = 0.05 ppm $°C^{-1}$ (Linear Technology claim a long-term stability of 2 μV per month for this temperature stabilized Zener type).

Great care has to be taken in the use of such devices as it is easy to generate thermoelectric voltages at the junctions of the leads which are much less stable than the Zener itself.

Zener diodes are two-terminal devices which have to be used in conjunction with a series resistor or a constant current supply.

7.2.2 Three-terminal Zener reference devices

Voltage references are now available which are basically a buried Zener followed by an op-amp. These give better stability than bandgap references, but still have the advantage of low output impedance and convenient output voltages. Examples are the industry standard devices REF01 for 10 V and REF02 for 5 V, which each have a temperature coefficient of 10 ppm $°C^{-1}$ maximum or 3 ppm $°C^{-1}$ for the best grade; they are available from several manufacturers.

7.2.3 Bandgap references

A transistor base–emitter junction voltage has a negative temperature coefficient, similar to a diode, but two transistors with different emitter current densities will have different base–emitter voltages:

$$V_{BE} \approx V_{BG} - CT$$

where V_{BG} is the bandgap voltage = 1.205, C is a parameter depending on the current density, and T is the absolute temperature.

For two transistors with different current densities

$$V_{BE1} \approx V_{BG} - C_1T \qquad (7.2)$$

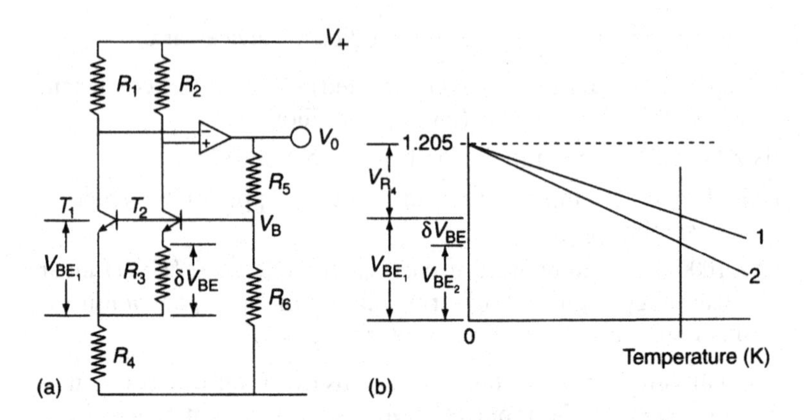

Figure 7.2 *Bandgap voltage reference: (a) circuit; (b) characteristics.*

$$V_{BE2} \approx V_{BG} - C_2T \tag{7.3}$$

Let $\delta V_{BE} = V_{BE1} - V_{BE2}$, then

$$\delta V_{BE} \approx (C_2 - C_1)T \tag{7.4}$$

Also, from (7.2), $V_{BE1} + C_1T \approx V_{BG}$.

Therefore,

$$V_{BE1} + \frac{\delta V_{BE}C_1}{(C_2 - C_1)} \approx V_{BG} = 1.205 \tag{7.5}$$

which is independent of T. So, if we could have a circuit which could develop this voltage, we would have a voltage source independent of temperature.

Consider the circuit shown in Figure 7.2 in which the collectors of T_1 and T_2 are held at the same potential by the open loop op-amp. Therefore,

$$i_1R_1 = i_2R_2 \tag{7.6}$$

$$\delta V_{BE} = i_2R_3 \tag{7.7}$$

The current through R_4 is

$$i_1 + i_2 = i_2\frac{(R_1 + R_2)}{R_1} \tag{7.8}$$

The voltage across R_4 is

$$V_{R_4} = R_4 i_2 \frac{(R_1 + R_2)}{R_1}$$

$$= R_4 \frac{\delta V_{\text{BE}}}{R_3} \frac{(R_1 + R_2)}{R_1} \tag{7.9}$$

From the circuit diagram we can see that

$$V_{\text{B}} = V_{\text{BE1}} + V_{R_4}$$

$$= V_{\text{BE1}} + R_4 \delta B_{\text{BE}} \frac{(R_1 + R_2)}{R_3 R_1} \tag{7.10}$$

The circuit is designed so that

$$\frac{R_4(R_1 + R_2)}{R_3 R_1} = \frac{C_1}{(C_2 - C_1)} \tag{7.11}$$

which makes V_{B} independent of temperature.

In practice, because of approximations in the calculations, V_{B} is nearer to 1.23 and not 1.205 as implied above. The exact value depends on the doping of the semiconductors. Thus output voltage is

$$V_{\text{out}} = V_{\text{B}} \frac{(R_5 + R_6)}{R_6} \tag{7.12}$$

More exact equations are given by Kuijk (1973) and other details by Sheingold (1986).

The output voltage of the complete circuit can be scaled to any convenient value and is taken from the output terminal of the op-amp, which is a low impedance point, so that the load regulation is much better. That is, within reason, the output voltage is not affected by the current drawn. For these reasons, a bandgap reference can be more convenient to use than a Zener which needs additional circuitry, but for the highest stability use a high quality buried Zener with a controlled temperature. The latest bandgap references approach Zener diodes in stability. For example the AD 780 operates at 2.5 or 3.0 V with a temperature coefficient of 3 ppm °C^{-1} and the best grades of AD 2710 and AD 2712 can achieve 10 V at 1 ppm °C^{-1}.

7.2.4 Ratiometric measurements

Sometimes the output of a sensor depends not only on the measurand but also on the voltage which powers the sensor. Examples are potentiometers for measuring position and d.c.-energized strain gauges. It then makes good sense to use the same voltage for the sensor as for the A/D converter, then changes in this voltage will not affect the output code of the A/D, and a larger temperature coefficient can be tolerated.

7.3 Bipolar binary codes

There are several ways of relating a digital **word** to the value it represents. These are **codes**. A binary digit is known as a **bit** and a word having N bits can represent any integral number from zero to $2^N - 1$, or from $-(2^N - 1)$ to $+(2^{N-1} - 1)$. The bits at the ends of the word are known as the **most significant bit** (MSB), and the **least significant bit** (LSB). In computing, the LSB is bit 0 and the MSB is bit N–1. Some converter manufacturers use the same terminology, but others call the LSB bit N and the MSB bit 1. To avoid confusion we shall use MSB and LSB wherever possible.

Straight binary code can represent positive numbers only, and to make it easier to read and write, a binary code is often expressed in groups of four bits. Each group can then be written as one hexadecimal (base 16) digit.

7.3.1 Offset binary

All zeros represents the maximum negative number and all ones represents the maximum positive number. Thus a negative number is indicated by the MSB = 0. This code is suitable for A/Ds and D/As because increasing analogue values are represented by increasing digital values throughout.

7.3.2 Two's complement

All zeros digital code represents zero analogue value, and all ones represents one LSB less. A negative number is indicated by the MSB = 1. This code is more suitable for arithmetic operations in computers. To convert offset binary to two's complement and vice versa, invert the MSB. Figure 7.3 compares these codes graphically, whereas Table 7.1 shows their numerical representation.

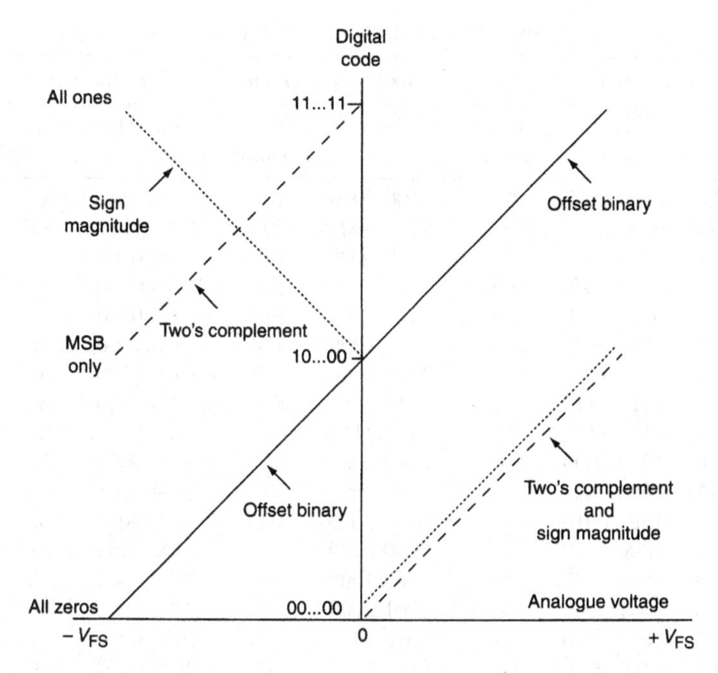

Figure 7.3 *Binary codes.*

7.3.3 Sign magnitude

The polarity of the value is represented by the MSB and magnitude (modulus) of the value is represented by the remaining bits. This code is not widely used because the discontinuity at zero makes it unnecessarily complicated.

7.3.4 Gray code

Consider what happens in practice to the binary code as the value is increased from one to two: as 0001 changes to 0010 the word may momentarily be 0011 or 0000, depending on which bit changes faster. In Gray code (Table 7.2) only one bit changes at a time so no incorrect readings are generated. Unfortunately the bits in this code are not uniformly weighted, so it is not widely used.

Figure 7.4 shows a simple circuit to convert binary code to Gray code and vice versa.

Table 7.1 *Comparison of binary codes*

Decimal	*Offset binary*		*Two's complement*		*Sign magnitude*	
	Binary	*Hexa-decimal*	*Binary*	*Hexa-decimal*	*Binary*	*Hexa-decimal*
–128	0000 0000	00	1000 0000	80	not available	
–127	0000 0001	01	1000 0001	81	1111 1111	FF
–64	0100 0000	40	1100 0000	C0	1100 0000	C0
–32	0110 0000	60	1110 0000	E0	1010 0000	A0
–16	0111 0000	70	1111 0000	F0	1001 0000	90
–8	0111 1000	78	1111 1000	F8	1000 1000	88
–4	0111 1100	7C	1111 1100	FD	1000 0100	84
–2	0111 1110	7E	1111 1110	FE	1000 0010	82
–1	0111 1111	7F	1111 1111	FF	1000 0001	81
0	1000 0000	80	0000 0000	00	0000 0000	00
+1	1000 0001	81	0000 0001	01	0000 0001	01
+2	1000 0010	82	0000 0010	02	0000 0010	02
+4	1000 0100	84	0000 0100	04	0000 0100	04
+8	1000 1000	88	0000 1000	08	0000 1000	08
+16	1001 0000	90	0001 0000	10	0001 0000	10
+32	1010 0000	A0	0010 0000	20	0010 0000	20
+64	1100 0000	C0	0100 0000	40	0100 0000	40
+127	1111 1111	FF	0111 1111	7F	0111 1111	7F
+128	not available		not available		not available	

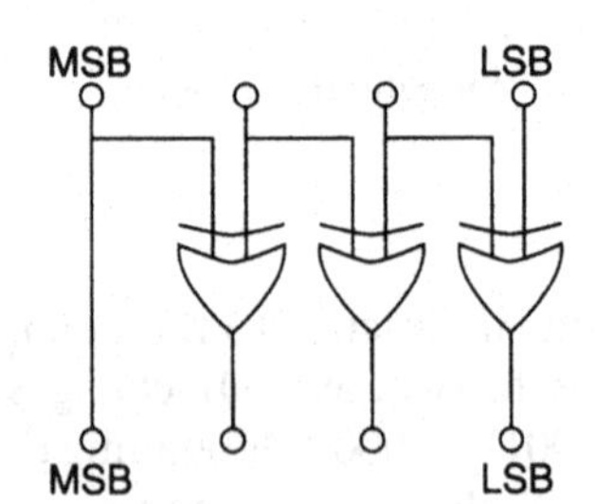

Figure 7.4 *Circuit to convert Gray code to binary and vice versa.*

Table 7.2 *Gray code*

Decimal	*Binary*	*Gray*	*Comment*
0	0000	0000	
1	0001	0001	bit 0 (LSB) changes
2	0010	0011	bit 1 changes
3	0011	0010	
4	0100	0110	bit 2 changes
5	0101	0111	
6	0110	0101	
7	0111	0100	
8	1000	1100	bit 3 (MSB) changes
9	1001	1101	
10	1010	1111	
11	1011	1110	
12	1100	1010	
13	1101	1011	
14	1110	1001	
15	1111	1000	

7.3.5 Binary coded decimal

Having 10 fingers and 10 toes, people have developed counting systems based on ten. Binary coded decimal (BCD) is a compromise in which each decimal digit is separately coded into a 4-bit binary word. Thus, an instrument reading of '1 2 3 4' would be coded 0001 0010 0011 0100.

7.3.6 American Standard Code for Information Interchange

This code is commonly known as ASCII, and uses seven or eight bits to represent punctuation, upper and lower case letters, numbers and control codes, known collectively as **characters**. ASCII is used to transmit data serially, i.e. one character at a time. ASCII codes are used to transmit data over the IEEE-488 bus and are listed in Table 11.5.

Seven bits gives $2^7 = 128$ different codes which is sufficient to represent all the numbers, upper and lower case letters, punctuation and control codes. The eighth bit (MSB) may be used to check parity. Parity checking is a simple form of error detection. If there are an odd number of ones in the 7-bit code, the MSB is set to 1; otherwise it is a 0. Thus there are always an even number of 1s in the 8-bit code. If, after processing or transmission, the 8-bit code is found to contain an odd number of 1s, an error has occurred. Alternatively the eighth bit may be used to extend the code for Greek letters, accented letters and mathematical or currency symbols. Several interpretations of codes 128 to 255 are in use with computers and are known as code pages.

7.4 Examples

The following examples show how to convert between the codes described above.

Example 7.1

A 10-bit A/D converter incorporates a bandgap voltage reference having a voltage of 1.235 V with a temperature coefficient of 100 ppm $°C^{-1}$.

1. Calculate the maximum temperature variation of the voltage reference which can be tolerated without causing more than 1 least significant bit error at maximum input.
2. If the converter is unipolar, calculate the output code corresponding to an input of + 0.500 V in binary and hexadecimal codes.
3. If the converter is bipolar, calculate the output code corresponding to inputs of + 0.500 V and –0.500 V in offset binary and two's complement codes.

Solution

1. LSB error at maximum input is 1 in 1023, i.e. 977.5 parts per million (ppm); temperature coefficient is 100 ppm K^{-1}. Therefore, maximum tolerable temperature change = 9.775 K. As temperature changes on the Celsius scale are the same as on the absolute (Kelvin) scale, this result may also be written as 9.8 °C.
2. The range of a unipolar converter is 0 to 1.235 – 0.0012 = 1.2338 V.

$$500\text{ mV} = \frac{500}{1.206} = 414.6\text{ LSB} = 415\text{ to nearest LSB}$$

To convert 415 to binary, divide by two:

Half 415 = 207 remainder = 1 = LSB
207 = 103 remainder = 1
103 = 51 remainder = 1
51 = 25 remainder = 1
25 = 12 remainder = 1
12 = 6 remainder = 0
6 = 3 remainder = 0
3 = 1 remainder = 1
1 = 0 remainder = 1

Therefore, 10-bit binary code for 500 mV is 01 1001 1111 and the hexadecimal code for 500 mV is 1 9 F.

3. The range of a bipolar converter is from –1.235/2 to + 1.235/2 – 0.0012, i.e. from –0.6175 V to +0.6163 V.

   In offset binary code,

   00 0000 0000 ≡ –0.6175 V
   10 0000 0000 ≡ 0.0000 V, i.e. analogue zero

   and 500 mV above zero is

   11 1001 1111 ≡ 0.5000 V

   In hexadecimal code this is 3 9 F.

   500 mV below zero = (0.6175 – 0.500) = 0.1175 V above minimum.

$$0.1175 \text{ V} = \frac{117.5}{1.206} = 97.43 \text{ LSB} = 97 \text{ to nearest LSB}$$

   To convert 97 to binary code, divide by two:

   Half 97 = 48 remainder = 1 = LSB
   48 = 24 remainder = 0
   24 = 12 remainder = 0
   12 = 6 remainder = 0
   6 = 3 remainder = 0
   3 = 1 remainder = 1
   1 = 0 remainder = 1

The 10-bit offset binary code for –500 mV is 00 0110 0001, and the hexadecimal code is 0 6 1.

In two's complement code,

00 0000 0000 ≡ 0.0000 V

01 1001 1111 ≡ 0.500 as in unipolar binary

For negative values, two's complement is the same as offset binary except that the MSB is inverted. From the above, we have

–500 mV ≡ 00 0110 0001 in offset binary

≡ 10 0110 0001 in two's complement

≡ 2 6 1 in hexadecimal

Example 7.2

Express the number 1001 0100 0101 in decimal and hexadecimal codes. Four 12-bit D/A converters each have a 10 V reference and negligible errors. What will the outputs be if the input to each converter is 1001 0100 0101 and they work in the following codes:

1. unipolar binary;
2. offset binary;
3. two's complement binary;
4. binary coded decimal.

Express the answers to an appropriate number of figures for the resolution of the converter.

Solution

Binary	1001	0100	0101	
Hexadecimal	9	4	5	by inspection
Decimal	(256×9) +	(16×4)	+ 5	
	= 2304 +	64	+ 5	= 2373

1. In 12-bit, unipolar binary code and with a 10 V reference,

$$1 \text{ LSB} = \frac{10}{2^{12}} = 0.002\,441\,4 \text{ V}$$

It is appropriate to calculate output voltages to the nearest millivolt, but do not round the value of the LSB before multiplication.

Binary code 1001 0100 0101 = 2048 + 256 + 64 + 4 + 1 = 2373 decimal. Thus, code 2373 gives

$$2373 \times 0.002\,441\,4 = 5.793\ 442\ 2\ \text{V}$$

and not $2373 \times 0.0024 = 5.6952$.

2. The same code applied to a 12-bit converter working in offset binary code would give an output of

$$-5.000 + 5.793 = 0.793$$

3. In two's complement binary, the MSB is called the sign bit. If it is 1, the value is negative, so in two's complement the same code would give an output of

$$5.793 - 10 = -4.207\ \text{V}$$

Alternatively, the value may be calculated by inverting the MSB and calculating the voltage as from offset binary code:

$$0001\ 0100\ 0101 = (256 \times 1) + (16 \times 4) + 5 = 325\ \text{decimal}$$

$$325 \times 0.002\,441\,4 = 0.793\ 455\ \text{V}$$

In offset binary code, this would give

$$-5 + 0.793\,455 = -4.206\ 545\ \text{V}$$

In summary, to the nearest millivolt, the code 1001 0100 0101 gives

5.793 V in unipolar binary code
0.793 V in offset binary code
–4.207 V in two's complement code

Example 7.3
Express 2684 decimal in hexadecimal and binary codes.

Solution
To convert a decimal number to binary, divide by two and note the remainder. Repeat the process until only 1 is left:

Decimal	*Remainder*	*Hexadecimal*
2684	0	
1342	0	C
671	1	
335	1	
167	1	
83	1	7
41	1	
20	0	
10	0	
5	1	A
2	0	
1	1	

Reading the remainder column from bottom to top gives the binary number, 1010 0111 1100.

Taking groups of four digits gives the hexadecimal code, A 7 C.

CHAPTER 8

Digital to analogue converters

8.1 Design of digital to analogue converters

So many good, inexpensive digital to analogue converters (D/As) are available as integrated circuits that nobody would build one from its component parts. Nevertheless, an applications engineer still needs to understand the design in order to make the best use of the ICs. There are three types:

- weighted resistor;
- ladder network;
- time division.

The ladder type is by far the most common of these three. The essential parts of a weighted resistor or ladder D/A converter are:

- a reference voltage;
- a set of resistors;
- a set of switches, one for each bit;
- a summing device to add together the contributions from all the bits.

8.1.1 Weighted resistor circuit

The resistors may be weighted according to any single-valued code; that is offset or two's complement binary but not Gray. The circuit is satisfactory up to approximately six bits, but at higher resolutions the large resistances are too large relative to leakages and/or the small resistances are too small relative to the ON resistances of the switches and so the circuit is not accurate (Figure 8.1).

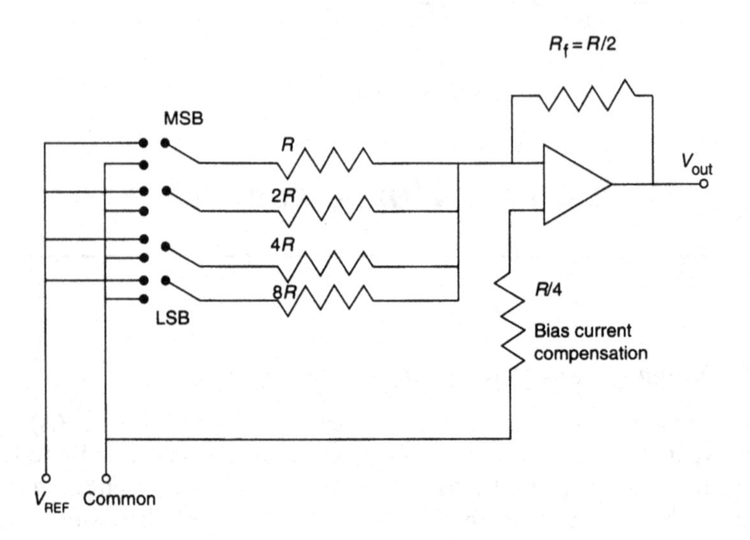

Figure 8.1 *Weighted resistor digital to analogue converter.*

8.1.2 Ladder networks

Improved results may be obtained by using a ladder network. All the resistances have the value R or $2R$, and R can be selected so that it is much less than the leakage but much more than the ON resistances of the switches. Integrated circuits can be designed so that the ratio of two (or more) resistors is known to a much higher accuracy than the value of either of the resistances separately. Figures 8.2(a) and (c) show two versions of the R–$2R$ ladder network. The reference has to be a low-impedance source so that the reference voltage does not change as the code changes. In both versions it can be seen that the resistance to earth from each of the nodes A, B, C and D is $2R$, regardless of the positions of the switches. Figure 8.2(a) illustrates the ladder in the voltage mode which works into a high impedance and generates a voltage:

$$V_{out} = \frac{V_{REF} D}{2^N} \tag{8.1}$$

where D is the digital code set on the switches and N is the number of bits.

To see how the voltage mode operates, consider Figure 8.2(b) which shows the equivalent network when the most significant bit

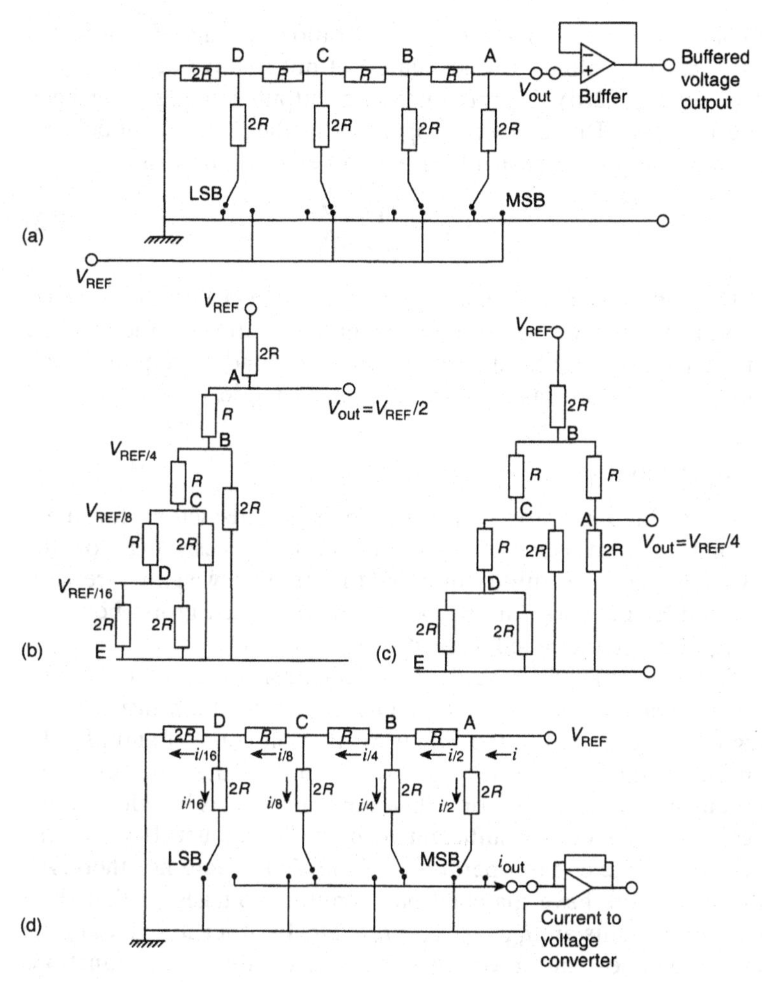

Figure 8.2 *Ladder type of digital to analogue converter: (a) voltage mode; (b) MSB only ON; (c) next bit to MSB ON; (d) current mode.*

only is on. It is easy to see that the output is ½V_{REF}. For a full analysis, use Thévenin's theorem.

If more than one bit is ON at the same time, the output is the sum of the voltages due to each separate bit. This result is given by the superposition theorem. In practice, the small, though finite, resistances of the switches and the reference voltage source cause small errors. The load resistance, i.e. from node A to ground, affects all bits equally and does not affect the ratio of voltages for different bits. That is, it affects the absolute but not relative accuracy.

In Figure 8.2(d) the reference and output voltages have been interchanged. This converter operates in the 'current mode' and gives an output current into a low impedance point of

$$i_{\text{out}} = \frac{V_{\text{REF}}\, D}{R \times 2^N} \tag{8.2}$$

This type is also known as current steering because the switches do not change the current drawn from the reference – they merely direct it into the output terminal or to ground as required. This mode is therefore faster than the voltage mode.

8.1.3 Quads

The weighted resistor configuration is unsatisfactory for high-resolution converters because of the finite resistance of the switches, but it is quite common to combine weighted resistor and ladder networks in groups of four bits, each group being an integrated circuit called a **quad**.

Figure 8.3(a) shows the details of a current-switched quad. The four Zener diodes couple the digital inputs to the transistors – a high logic level makes the Zener for that bit conduct and lifts the emitter voltage above the common voltage of the bases and cuts off current in the transistor. When the logic level is low the voltage across the Zener is insufficient to make it conduct. It is reverse biased and the circuit operates as if the Zener were not there. All the transistors have matched base–emitter voltages so that their emitter currents are i_{REF}, $\frac{1}{2}i_{\text{REF}}$, $\frac{1}{4}i_{\text{REF}}$ etc. In order to maintain the matched base–emitter voltages, the size of the emitter junctions is made proportional to the current.

Figure 8.3(b) shows how the current from one quad is divided by 16 before being summed with the next more significant quad. Figure 8.3(c) shows two interquad dividers giving a total resolution of 12 bits. Note the modification in the resistor value from R to $\frac{16}{15}R$ because it is shunted by the $(R + 15R)$ from the previous stage.

8.1.4 Timesharing D/As

Referring to Figure 8.4, the counter counts continuously. When the count is all 0s, the set input of the bistable pulses high and

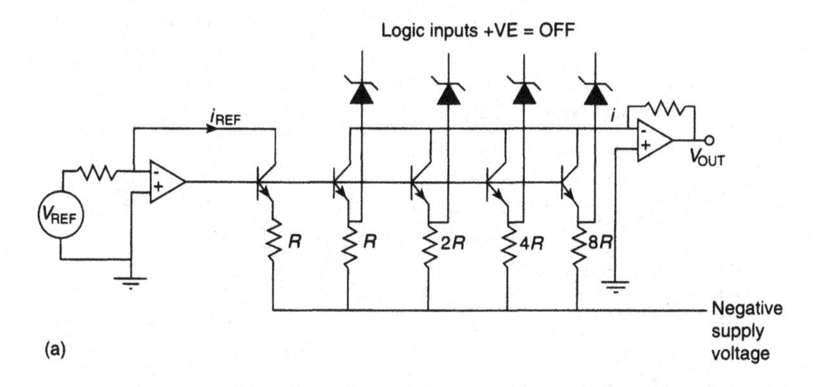

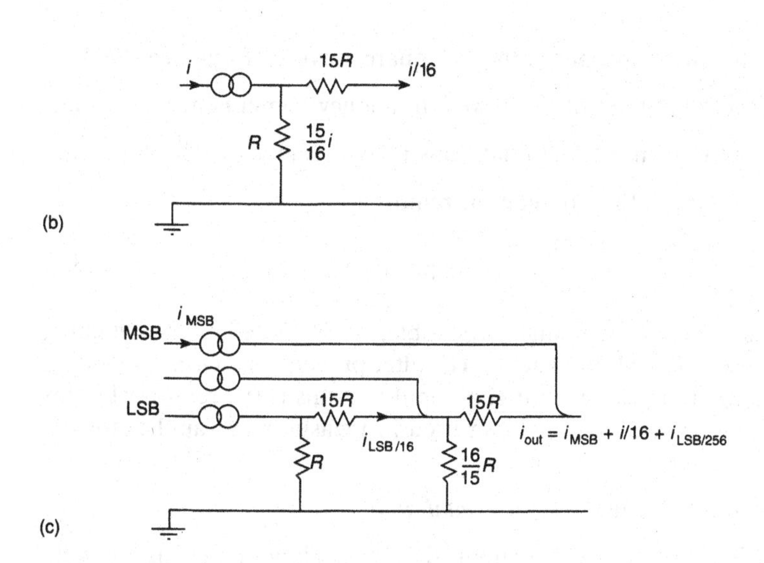

Figure 8.3 *Quads: (a) current-switched quad; (b) interquad divider by 16; (c) interquad dividers by 16 and by 256.*

sets Q to 1; when the count matches the digital input, the comparator goes high and resets the bistable ($Q = 0$). The bistable outputs drive CMOS switches which connect the analogue output terminal either to V_{REF} or to earth. To obtain a steady d.c. output is necessary to follow the switches by a low-pass filter. To estimate the required break frequency of the filter, consider what happens when only the MSB is on (see equation 10.13):

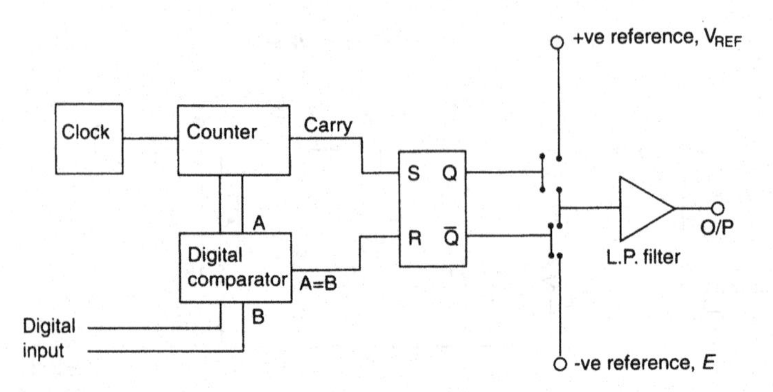

Figure 8.4 *Timesharing type of digital to analogue converter.*

the switch output is then a square wave of frequency $f_c/2^N$

the amplitude of the lowest frequency component is $V_{REF}.2/\pi$

the filter must attentuate this to less than 1 LSB = $V_{REF}/2^N$

therefore, the attenuation required is

$$\frac{4 \times 2^N}{\pi} \text{ at a frequency of } f_c/2^N \tag{8.3}$$

Thus f_c should be as high as possible, and is limited by the switching time of the CMOS switches. The filter prevents the D/A responding rapidly to a change in digital input and this is the reason why this type of converter is seldom used in measurement applications.

8.1.5 D/A converters for digital audio

A development of the circuit described above is the sigma–delta D/A used in digital audio systems (Analog Devices 1989–90). The principle is similar, with two important modifications. First, the incoming digital signal is interpolated so that the sampling frequency of the actual converter is higher than the sampling frequency of the audio signal, often times 128. Secondly, a high-order digital filter is used after the one-bit converter in place of a counter. This filter provides a running average of the pulse train, which obviates the need for a long counting time even for high resolution converters.

8.2 Bipolar D/As

The circuits described up to now give zero analogue output for zero digital input, i.e. straight binary. To make an offset binary converter, a fixed current worth one MSB negative is fed into the summing junction so that the output is maximum negative for zero code.

8.3 Multiplying D/As

Some D/As are designed to use a variable analogue input in place of a fixed reference voltage. These are multiplying converters and may be used as digitally controlled potentiometers for variable-gain amplifiers, filters, voltage-controlled oscillators and so on. If both digital and analogue inputs are bipolar, the converter is described as **four quadrant**.

8.4 Accuracy of D/A converters

Errors in D/A converters are illustrated in Figure 8.5 which shows the relationships between the digital input code (V_D) and the analogue output voltage V_A. The thin line represents the ideal and the thick line the reality.

Absolute and relative accuracy

Absolute accuracy is specified by the worst-case error in output expressed in volts, relative to an absolute, external standard. Relative accuracy is specified by the worst-case error expressed as a proportion or percentage of the full scale.

Thus variations in the reference voltage affect the absolute accuracy, but not the relative accuracy. In many applications relative accuracy is more important than absolute accuracy.

Gain error

The output which would be obtained for a given input code if there were no errors is denoted the **nominal output** and is represented by the thin lines in Figure 8.5. A difference between actual and nominal output which is proportional to the nominal output is shown in Figure 8.5(a). This error expressed as a fraction of the nominal output is the gain error.

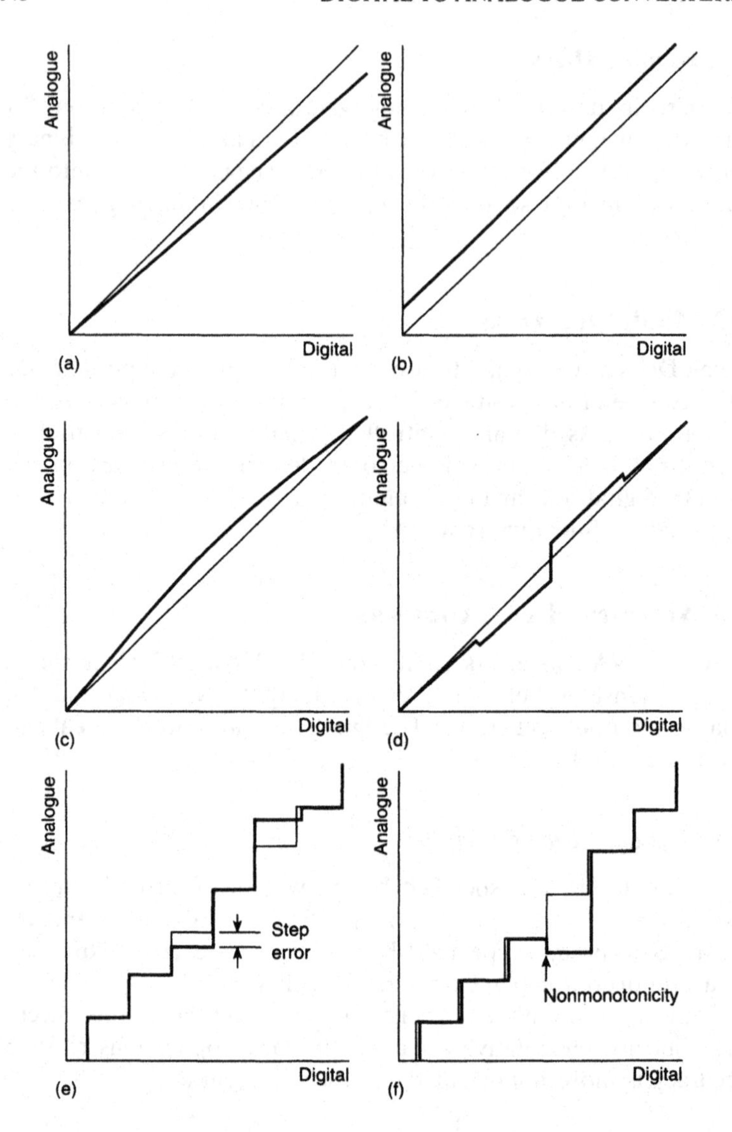

Figure 8.5 *Errors in digital to analogue converters: (a) gain error; (b) offset error; (c) terminal nonlinearity; (d) terminal nonlinearity in a switched resistor D/A; (e) differential linearity error; (f) nonmonotonicity.*

The largest input code which can be applied to the converter is all ones, which has the value $2^N - 1$ and the corresponding output is the **full-scale voltage**, designated V_{FS}. In a bipolar converter, the **full-scale range** is the full analogue range from maximum negative to maximum positive value, i.e. $V_{FR} = V_{+FS} - V_{-FS}$ (Gordon, 1978; Clayton, 1982). Thus, for a unipolar converter, the average measured value of 1 LSB is

$$\mathrm{LSB}_{av} = \frac{V_{FS} - V_{OS}}{2^N - 1} \tag{8.4}$$

where V_{OS} is the offset voltage.

For a bipolar converter

$$\mathrm{LSB}_{av} = \frac{V_{+FS} - V_{-FS}}{2^N - 1} = \frac{V_{FR}}{2^N - 1} \tag{8.5}$$

where V_{FR} is the full range voltage.

Some manufacturers define full-scale range as 'twice the value of the most significant bit' (Scheingold, 1986). It follows from this definition that

$$\mathrm{LSB}_{av} = \frac{V_{FR}}{2^N} \tag{8.6}$$

The nominal value of 1 LSB may be quoted in the specification or may be calculated as

$$\mathrm{LSB}_{nom} = \frac{V_R}{2^N} \tag{8.7}$$

where V_R is the nominal value of the reference voltage.

Variations in the actual reference voltage result in gain errors:

$$\text{gain error} = \frac{(V_{FS} - V_{OS})\, 2^N}{V_R(2^N - 1)} - 1 \tag{8.8}$$

Offset error

When the all zeros code is applied to a unipolar converter, the output should be zero. If it is not, the difference is the offset. The **offset** is defined as a difference between actual and nominal output which is independent of output and measured as the actual output for the digital code which ought to give zero output. In a unipolar or two's complement converter, this is all zeros, but in an offset binary converter it is MSB = 1 and all other bits = 0 (Figure 8.5(b)).

Linearity

If the output is not proportional to the input (after allowing for the offset), the difference is the nonlinearity, which may be quantified in several ways. Figure 8.5(c) shows the **terminal nonlinearity** which compares the actual output with that represented by a straight line joining the **terminal points** which correspond to minimum and maximum output. The largest discrepancy between these lines is the **terminal** or **integral linearity error**. In a time-sharing D/A, and in A/Ds using a counter, the nonlinearity is a smoothly changing function, as in Figure 8.5(c). However, in switched resistor D/As, and consequently in successive approximation A/Ds, the nonlinearity depends which bits are ON for any particular output and so the nonlinearity is a jagged function, as shown in Figure 8.5(d).

The output of a perfect D/A converter should, of course, increase in a series of small, equal steps as the code is increased. Figures 8.5(e) and (f) show a small part of the output/input characteristic. Each step represents one least significant bit (LSB). The maximum difference between an actual step and a nominal one is the **differential linearity error**, **step error**, or the **nonlinearity**. This is shown in Figure 8.5(e). If the output ever decreases as the input increases, the converter is said to be **nonmonotonic** (Figure 8.5(f)). This fault is rare, but renders a converter useless for use in a control loop. Obviously this can occur with the type of nonlinearity in Figure 8.5(d) but not as in (c), so converters with the smooth characteristic are sometimes advertised as inherently nonmonotonic.

Settling time

If the input code is suddenly changed, there is a small delay before the output reaches its final value. The **settling time** is defined as the time for the output to reach and stay within ±½ LSB of its final value. It is usually specified for a full-range change, but it may also be defined for the switching point of the MSB.

Glitches

When the input code is changed, there may be a transient in the output voltage caused by the difference in the turn-off and turn-on times of different bits. This is known as a glitch, and the largest one usually appears when the MSB changes ON at $\frac{1}{2}V_{FR}$. The next largest are at $\frac{1}{4}V_{FR}$ and $\frac{3}{4}V_{FR}$, and so on. Glitches may be removed by a low-pass filter if a fast response is not required. Alternatively, a sample and hold circuit (Chapter 5) may be used to hold the converter output until it has settled to its next value. The settling time for a converter includes the time for the glitch to disappear.

8.5 Sources of error

Reference voltage

An error in the reference voltage causes gain error; that is, it affects the absolute accuracy but not the relative accuracy. The temperature coefficient of gain error will be specified by the converter manufacturer but, if the reference voltage is external, its temperature coefficient will have to be considered as well.

Load resistance

Variation in the load resistance of an unbuffered resistance network also causes gain error.

Network resistors

Errors in the values of resistors in the ladder network cause nonlinearity.

Switches

The ON resistance of each switch effectively adds to the resistance with which it is in series.

Buffer

Nonlinearity may also occur in the output buffer.

Temperature

As resistances, particularly those of semiconductors, change with temperature, a complete specification must include not only typical and maximum values for the errors defined above but also **temperature coefficients** for gain, offset and linearity errors.

A further discussion of errors in D/As will be found in Young (1987).

8.6 Testing D/A converters

8.6.1 Input codes

The digital input code may be supplied by:

- manually operated switches;
- a digital counter;
- a computer

The analogue output may be measured by a digital voltmeter or an A/D converter of better accuracy than the one under test. It is not necessary to test all possible values of the input code. All bits OFF indicates offset, V_0; all bits ON indicates a full-scale or full-range output, V_{FS}. Each bit ON separately constitutes a bit scan or walking-ones test, and is used to measure terminal nonlinearity.

8.6.2 Step error

The **differential linearity error** or **step error** may be measured by turning several low bits ON and measuring the output, then turning these OFF and the next higher one ON. This is a bit transition which should ideally be one LSB. The worst error will usually be found as the MSB comes ON and all the other bits go OFF.

8.6.3 Dynamic testing of D/A converters

A reference D/A is required which has at least two bits better resolution and accuracy than the D/A under test. Both must have the same output range. The digital inputs are scanned together and the difference between their output voltages is measured (Figure 8.6(a)). To measure differential linearity, code D is applied to the test D/A and $D-1$ to the reference D/A. The difference in

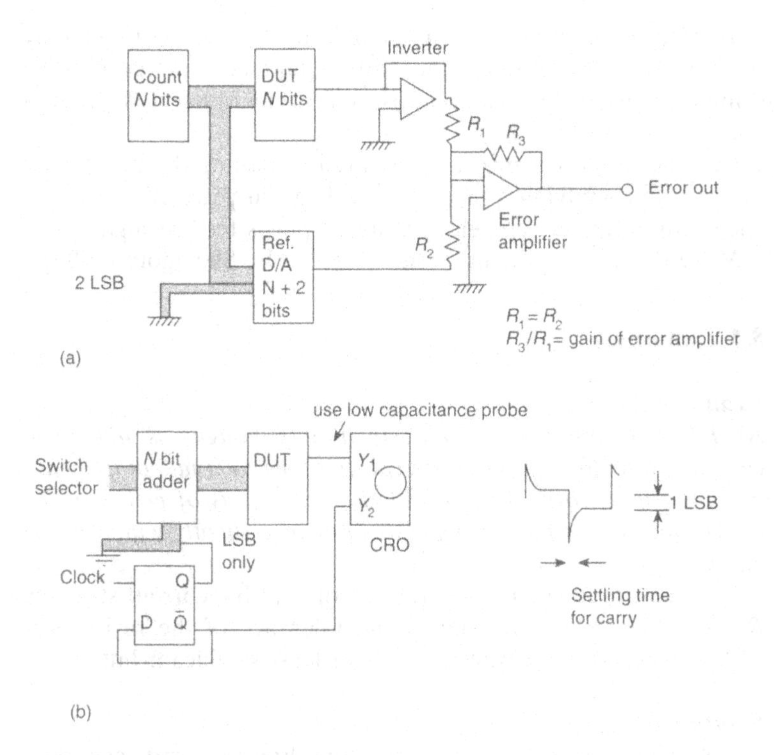

Figure 8.6 *Test circuits for D/A converters: (a) offset, gain and nonlinearity; (b) differential nonlinearity and settling time – to examine full-scale settling time connect all digital inputs of device under test (DUT) to Q.*

output should, of course, be 1 LSB. The input bits of the test device are then scanned. An alternative arrangement is shown in Figure 8.6(b). Any input code can be selected on the switches. The LSB is switched ON and OFF at clock frequency, so that the output should be a square wave in phase with the LSB. The output is monitored on a high-frequency CRO with a bandwidth at least 60 MHz using a properly adjusted, high-impedance probe. Suppose the switches on an 8-bit converter were set to 0111 1110 = 127, then the code applied to the A/D would alternate between 127 and 128. If the converter were nonmonotonic at this point, the output for 128 would be smaller than for 127. This would show as a phase reversal on the CRO. This test can be difficult as the noise level on the D/A output may be not much less than 1 LSB and therefore the signal-to-noise ratio of the difference signal may be poor.

To measure settling time the input code is switched repeatedly from all zeros to all ones. The same equipment may be used to examine glitches. For this purpose the input is scanned through the codes to be tested.

Feedthrough error may be measured by setting the input code to zero, and applying a sinusoidal voltage in place of the reference. Any a.c. which appears at the output is feedthrough.

More details of dynamic tests are given by Sheingold (1986).

8.7 Examples

Example 8.1

As D/A converters are almost invariably bought as integrated circuits, we seldom consider the values of the internal components. However, it is worthwhile considering the effects of errors in the ladder network as the resistances will change slightly with temperature.

Calculate the maximum step error of a 12-bit, current-steering, R–$2R$ ladder D/A converter, if the tolerance of the resistors is 0.01%. Express this result in terms of least significant bits.

Solution

The maximum step error or differential linearity error occurs at the transition when the most significant bit is switched ON and all the other bits are switched OFF.

The tolerance of the resistors is ± 0.01%, so the error is largest if the resistor for the MSB is 0.01% too large and the other resistors are 0.01% too small, or vice versa. The current then divides at the first node in the ratio 1.01:0.99 instead of 1:1, and the output voltage is 0.01% too small for the MSB and 0.01% too large for the other bits. This is a differential error of 0.02% of output. The output for the MSB for the converter in question is 5 V = 2048 LSB. Thus, the step error is

$$0.0002 \times 5 = 1 \text{ mV}$$

or

$$\begin{aligned} 0.0002 \times 2048 &= 0.4096 \text{ LSB} \\ &= 0.41 \text{ LSB to reasonable accuracy.} \end{aligned}$$

Example 8.2

The binary equivalent of 2684 is applied to a 12-bit 5 V converter. What would the output voltage be if the converter is working in (a) unipolar binary code and (b) offset binary code? Quote the answer to the appropriate number of significant figures.

Solution

One LSB in a 12-bit, 5 V converter is

$$5/2^{12} = 5/4096 = 0.001\ 220\ 7\ \text{V}$$

so it is reasonable to quote the output voltage to the nearest millivolt. In unipolar binary code, 2684 represents

$$5 \times 2684/4096 = 3.276\ \text{V}$$

In offset binary code, 2684 represents

$$5 \times 2684/4096 - 2.5 = 0.776\ \text{V}$$

Example 8.3

A 10-bit, unipolar D/A converter has a nominal output of 10.000 V with all bits ON. The analogue outputs were measured as 8.794 and 0.981 V, respectively, for digital input codes corresponding to decimal 900 and decimal 100. Calculate the offset and gain errors assuming perfect linearity.

The nonlinearity errors were measured separately and the bit errors were

bit 1 (MSB)	+ 0.6 LSB
bit 2	− 0.4 LSB
bit 3	+ 0.1 LSB
bits 4 to 10	negligible

Calculate the maximum differential linearity error and the input code at which it occurs, and also calculate the terminal linearity error.

Solution

Nominal output with all bits ON = 10.000 V.

$$\text{All bits on} = 2^{10} - 1 = 1023$$

Therefore,

$$\text{LSB} = \frac{10.000}{1023} = 9.775\ \text{mV}$$

Table 8.1 *D/A test results for Example 8.3*

	Code	*Calculated output*	*Measured output*
Decimal	900	8.7976 V	8.794 V
Decimal	100	0.9775 V	0.981 V
Difference	800	7.8201 V	7.813 V
LSB	1	9.775 mV	9.766 mV

Because we assume perfect linearity, we can calculate the gain error from the measured values (Table 8.1) of any two points:

$$\text{gain error} = \frac{9.766 - 9.775}{9.775} = 9.207 \times 10^{-4} = 0.092\%$$

Again assuming perfect linearity, when the input code is zero the measured value of the output will be

$$0.981 - \frac{(8.794 - 0.981)}{800} \times 100 = 0.004\,375 = 4.4\text{ mV}$$

Maximum step error occurs when the bit with the largest error comes on. That is at the transition to MSB only ON:

error at 01 1111 1111 = −0.4 + 0.1 = −0.3 LSB

error at 10 0000 0000 = +0.6 LSB

change = +0.9 LSB = differential linearity error

The largest linearity error occurs when bits 1 and 3 are ON together, i.e. error at code

10 1000 0000 = + 0.6 + 0.1 = 0.7 LSB

The terminal linearity error is

$$\frac{0.7}{1023} = 6.843 \times 10^{-4} = 0.068\% \text{ of range}$$

Example 8.4

How many digits would be required in a digital voltmeter to be used for testing a 10-bit D/A converter?

From the test results given in Table 8.2 for a 4-bit converter calculate the following parameters:

- least significant bit;
- offset voltage;
- gain error;
- terminal linearity (for most significant bit only);
- differential linearity (for each bit transition).

The reference voltage is 3.200 V.

Table 8.2 *Data for Example 8.4*

Input code	*Output (mV)*
0000	123
0001	313
0010	536
0011	727
0100	997
0111	1598
1000	1607
1111	3083

Solution

For a 10-bit converter, the resolution is 1 in $2^{10} = 1$ in 1024. Therefore, the dvm needs a resolution and a corresponding accuracy an order better; that is, 1 in 10 000. This is known as a 4½-digit dvm because it has four digits which can have any value from 0 to 9 and a 'half digit' which can have the values 0 or 1 only.

$$\text{Nominal value of LSB} = \frac{3200}{2^4}$$

$$= 200 \text{ mV, designated LSB (nominal)}$$

Terminal points:

$$0000 = 123 \text{ mV, the offset}$$

$$1111 = 3083 \text{ mV, the full range}$$

Full range output swing = 3083−123 = 2960 mV.

Measured average LSB is

$$\frac{3083 - 123}{15} = 197.3 \text{ mV, designated LSB (measured)}$$

The gain error is

$$\frac{(3083 - 123) \times 16}{15 \times 3200} - 1 = -0.0133 = -1.3\%$$

The calculated value of output with only MSB on is

$$\frac{(3083 - 123) \times 8}{15} + 123 = 1702 \text{ mV}$$

Measured value of output with only MSB on is 1607 mV.

Terminal linearity error with only MSB on is –95 mV

$$= \frac{95 \times 100}{(3083 - 123)} = 3.2\% \text{ of full range voltage change}$$

The differential linearity error may be calculated by compiling Table 8.3 using the test results given in Table 8.2.

Table 8.3 *Differential linearity test results for Example 8.4*

Input code change		*Output change*	*Output change – 1 LSB*	*Error*
To	*From*	*(mV)*	*(mV)*	*(mV)*
0001	0000	313 – 123 = 190	190 – 197.3	–7.3
0010	0001	536 – 313 = 223	223 – 197.3	+25.7
0100	0011	997 – 727 = 270	270 – 197.3	+72.7
1000	0111	1607 – 1598 = 9	9 – 197.3	–188.3

As usual, the largest error occurs at the MSB transition. When the calculations have been completed, the answer is rounded to a precision compatible with the original data, which is to the nearest millivolt.

Thus, the differential linearity error is –188 mV.

Example 8.5

In developing a circuit, we sometimes get unexpected results and have to work backwards to find the cause.

A test of an 8-bit D/A converter gives the results below:

Bit no.	*Digital code*	*Measured bit error (mV)*
1	128	−1.11
2	64	+0.40
3	32	−0.26
4	16	−80.91
5	8	+81.44
6	4	+0.11
7	2	+0.18
8	1	+0.15

The output for all bits ON is 2580.15 mV, and the offset is + 2.10 mV. Calculate the correct and measured outputs for each bit and hence the terminal nonlinearity. Comment on the monotonicity of this converter.

From the above data, there appears to be a fault on the digital side. Deduce what this is. If it were corrected, what would the bit errors and the terminal nonlinearity become?

Solution

Table 8.4 completes the test results given above. Column 4 is calculated as follows:

$$\text{calculated output voltage} = \frac{(V_{FS} - V_{OS})\,D}{255} + V_{OS}$$

$$= \frac{(2580.15 - 2.10)\,D}{255} + 2.10$$

$$= 10.1100D + 2.10$$

where D represents the digital code.

The measured output (column 5) is not given in the question, but may be deduced by adding the measured bit error to the calculated voltage. Comparing the measured and calculated outputs shows that when bit 4 is ON, the output is what would be expected from bit 5, and vice versa. We deduce that the wires to bits 4 and 5 have been crossed. If the wiring were corrected, then the errors would be:

bit 4: 82.95 − 82.98 = −0.03 mV

bit 5: 164.42 − 163.86 = +0.56 mV

Table 8.4 *D/A test results for Example 8.5*

Bit no.	*Digital code*	*Measured bit error (mV)*	*Calculated output (mV)*	*Measured output (mV)*	*Error after bits 4 and 5 interchanged*
1 MSB	128	–1.11	1296.18	1295.07	–1.11
2	64	+0.40	649.14	649.54	+0.40
3	32	–0.26	325.62	325.36	–0.26
4	16	–80.91	163.86	82.95	+0.56
5	8	+81.44	82.98	164.42	–0.03
6	4	+0.11	42.54	42.65	+0.11
7	2	+0.18	22.32	22.50	+0.18
8	1	+0.15	12.21	12.36	+0.15

The terminal linearity error is the sum of all the positive bit errors which theoretically is equal to the sum of the negative bit errors. (In practice these usually differ slightly due to temperature variations in the chip as different bits are ON.) In the uncorrected state:

the sum of the positive errors

$= 0.40 + 81.44 + 0.11 + 0.18 + 0.15$

$= 82.28$

the sum of the negative errors

$= 1.11 + 0.26 + 80.91$

$= 82.28$

The converter is nonmonotonic as the input codes goes from 15 to 16. After correction,

the sum of the positive errors

$= 0.40 + 0.56 + 0.11 + 0.18 + 0.15$

$= 1.4$

the sum of the negative errors

$= 1.11 + 0.26 + 0.03$

$= 1.4$

Therefore the terminal linearity error = 1.4 mV.

The converter is now monotonic.

Example 8.6

This example illustrates how a simple fault can lead to some very odd results.

The wiring to an 8-bit digital to analogue converter has been damaged so that bit 6 (= 4 LSB) is always ON and bit 7 (= 2 LSB) is always OFF. All the other bits are correct. Calculate:

- the offset;
- the gain error;
- the terminal nonlinearity;
- the differential nonlinearity.

Solution

The question defines the weight of bit 6 as 4 LSB and of bit 7 as 2 LSB, so we can deduce the weights of all the bits:

						ON	*OFF*	
Bit no.	1	2	3	4	5	6	7	8
Weight	128	64	32	16	8	4	2	1

When all bits should be OFF, 6 is ON and the output is 4 LSB, which is the offset.

When all bits should be ON, 7 is OFF and the output is 255 – 2 = 253.

$$\text{gain error} = \frac{253 - 4}{255} - 1 = -2.35\%$$

The maximum error occurs when bit 6 should be OFF; error = 4 LSB.

$$\text{terminal linearity error} = \frac{4}{253 - 4} = 1.61\%$$

The maximum differential linearity error occurs when error changes from –2 to + 4, i.e.

$$6 \text{ LSB} = 2.41\% \text{ of the full range}$$

Table 8.5 *Differential linearity test results for Example 8.6*

	Correct I/P	*Actual input*	*Error*	*Step error*
0	0000 0000	0000 0100 = 4	+ 4	end point
				0
1	0000 0001	0000 0101 = 5	+ 4	
				–2
2	0000 0010	0000 0100 = 4	+ 2	
				0
3	0000 0011	0000 0101 = 5	+ 2	
				–2
4	0000 0100	0000 0100 = 4	0	
				0
5	0000 0101	0000 0101 = 5	0	
				–2
6	0000 0110	0000 0100 = 4	–2	
				0
7	0000 0111	0000 0101 = 5	–2	
				6 MDLE*
8	0000 1000	0000 1100 = 12	+ 4	
				0
9	0000 1001	0000 1101 = 13	+ 4	= MTLE†
⋮	⋮	⋮		
255	1111 1111	1111 1101 = 253	–2	end point

*MDLE = maximum differential linearity error.
†MTLE = maximum terminal linearity error.

An alternative solution is by compiling Table 8.5. This is a more tedious, but more reliable method of arriving at the same result. The distance between end points should be 255, but is actually 249 LSB. The jump of 7 LSB from 7 to 8 will be repeated from 15 to 16, from 23 to 24, from 31 to 32 and so on.

CHAPTER 9

Analogue to digital converters

9.1 Early designs

In order to understand how modern designs of analogue to digital converters (A/D) work it is worth looking at some obsolete designs (Figure 9.1) to see where errors originated and how modern designs evolved.

The single ramp type is very simple. The 'start' pulse sets the bistable which opens the clock gate and starts the counter. At the same time S_1 is closed and applies the voltage to be measured to an integrator circuit. When the output of the integrator equals the reference voltage, the comparator resets the bistable and stops the counter. The count is then a measure of V_{IN}.

Errors are introduced by the offset in the comparator and delays in switching. It is also necessary that the clock frequency and the integrator time constant are both constant and accurate. The most serious weakness of this design is that the count is actually a measure of the mean value of V_{IN} over the integration period and this period is not fixed but depends on the value of V_{IN}.

The counter type achieves perfect synchronization of the counter and the ramp by using a D/A converter driven by the counter to produce a staircase waveform and comparing it with V_{IN}.

The count, when the staircase equals the input, measures the value of the input – at that moment. The **conversion rate** of a counter type is very low because the counter is reset to zero every time. The GEC Plessey ZN425 includes a counter and an 8-bit D/A, and may be used in this way.

If the counter can count both up and down, then the output of the D/A can be made to follow or 'track' V_{IN} if it changes, provided that the change is not too rapid. The limit is set by the **slew rate**,

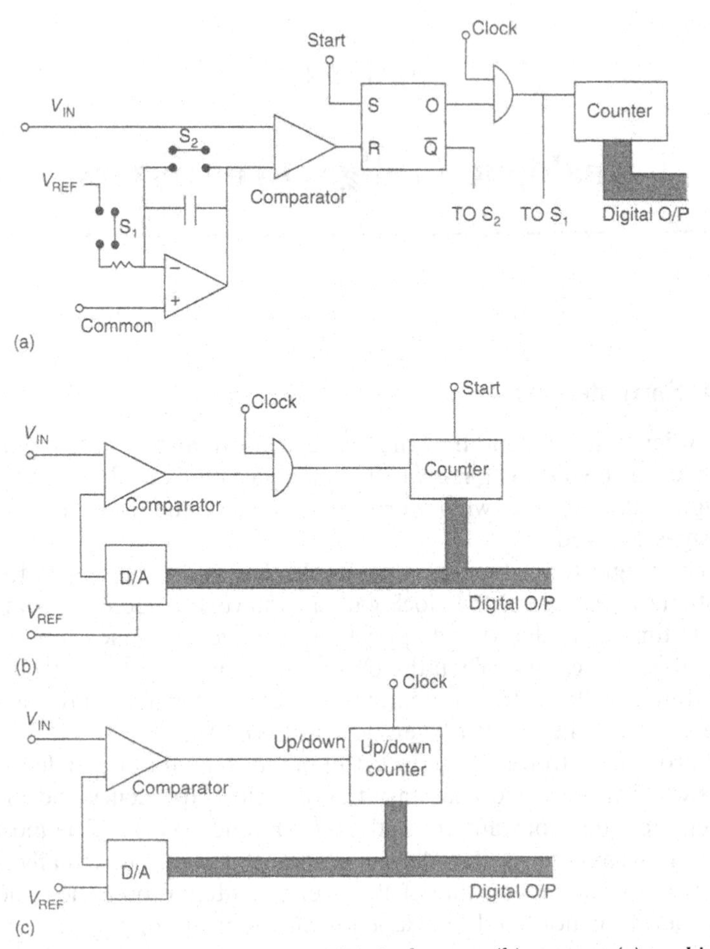

Figure 9.1 *Simple A/D converters: (a) single ramp; (b) counter; (c) tracking.*

or maximum rate of change of voltage at the output of the D/A which is 1 LSB per clock pulse, i.e. $f_c V_{REF}/2^N$.

Now consider a sinusoidal input signal whose peak-to-peak amplitude equals the range of the converter:

$$v = \frac{1}{2} V_{REF} \sin(2\pi f_s t)$$

$$\frac{dv}{dt} = \frac{1}{2} V_{REF} \times 2\pi f_s \cos(2\pi f_s t)$$

$$\max \frac{dv}{dt} = V_{REF}\pi f_s$$

This is the maximum slew rate of the signal and, for accurate conversion, it must not exceed the slew rate of the converter:

$$V_{REF}\pi f_s \leqslant V_{REF} f_c 2^N$$

$$f_s \leqslant \frac{f_c}{\pi \times 2^N} \tag{9.1}$$

If we substitute a few typical figures, we can see that perfect tracking is limited to relatively low frequencies, which is why this design is seldom used now. For example, if f_c is 1 MHz and N is 10 then,

$$\text{maximum } f_s = \frac{10^6}{1024\pi} = 310.8 \text{ Hz}$$

9.2 Integrating and nonintegrating converters

If the input voltage to a converter is not absolutely constant (and it hardly ever is), there are two measurements which are worth making. One is the instantaneous voltage at a known moment, which requires a fast converter; the other is the mean voltage over a specified interval, which is what an integrating converter measures. The mean voltage over an unknown or variable interval is not a useful measurement.

Integrating converters include:

- dual ramp;
- triple ramp and multislope;
- charge balance, voltage to frequency with counter and delta–sigma (Δ–Σ).

Nonintegrating converters include:

- successive approximation;
- flash.

9.3 Integrating converters

Consider what happens when the input (which should be steady) varies during the integration time (τ). Let the input V_{IN} be

$$V_{IN} = V + A \sin(2\pi f t + \phi) = V + A \sin(\omega t + \phi)$$

where V is the voltage to be measured, $A \sin(\omega t + \phi)$ is an unwanted ripple superimposed on it (e.g. mains hum), t is measured from the start of the integration period, and ϕ is the phase angle of the ripple at $t = 0$.

The voltage $A \sin(\omega t + \phi)$ is described as series mode noise or interference or, in American usage, normal mode.

The digital output of the converter D is given by:

$$D = \frac{1}{\tau}\int_0^{\tau} [V + A \sin(\omega t + \phi)]\, dt$$

$$= \frac{1}{\tau}\left[V_T - \frac{A \cos(\omega t + \phi)}{\omega}\right]_0^{\tau}$$

$$= \frac{1}{\tau}\left[V_T - \frac{A\{\cos(\omega t + \phi) - \cos\phi\}}{\omega}\right]$$

$$= \frac{1}{\tau}\left[V_T + \frac{A\{2 \sin\frac{1}{2}(\omega\tau + 2\phi) \sin\frac{1}{2}(\omega\tau)\}}{\omega}\right]$$

$$= V + \frac{A\{2 \sin(\frac{1}{2}\omega\tau + \phi) \sin(\frac{1}{2}\omega\tau)\}}{\omega\tau}$$

The error is

$$D - V = \frac{A\{2 \sin(\frac{1}{2}\omega\tau + \phi) \sin(\frac{1}{2}\omega\tau)\}}{\omega\tau} \quad (9.2)$$

The error is zero when $\frac{1}{2}\omega\tau = 0$ or $k\pi$, where k is any integer, i.e. when $\omega\tau = 2k\pi$ or $f = k/\tau$. This is true for all values of ϕ. Thus, the noise does not cause any error if the integration time τ is a multiple of the period of the noise, and the phase of the noise is irrelevant.

However, when $\frac{1}{2}\omega\tau = 0$ or $k\pi$ is not true, the error varies with ϕ and each conversion gives a different result.

The maximum error occurs when $\sin(\frac{1}{2}\omega\tau + \phi) = \pm 1$:

$$E_{max} = \pm \frac{A \sin(\frac{1}{2}\omega\tau)}{\frac{1}{2}\omega\tau} \quad (9.3)$$

An alternative method of analysing this error is to consider that the converter 'sees' the noise through a window in time of duration τ and to compute the convolution function.

9.3.1 Series mode rejection ratio

The series mode rejection ratio (SMRR) is defined as

$$\text{SMRR} = \frac{\text{maximum noise input}}{\text{maximum error output}}$$

When the noise waveform is sinusoidal, this becomes

$$\text{SMRR} = \frac{\text{peak noise}}{\text{peak error}}$$

$$= \frac{A}{\{A \sin(\frac{1}{2}\omega\tau)\}/(\frac{1}{2}\omega\tau)}$$

$$= \frac{\frac{1}{2}\omega\tau}{\sin(\frac{1}{2}\omega\tau)} \qquad (9.4)$$

An alternative way of arriving at this equation is to consider that the noise signal, $A \sin \omega t$ is observed through a window of duration τ. The resulting error is given by the convolution integral. The SMRR is usually quoted in decibels:

$$\text{SMRR} = 20 \log_{10} \frac{\frac{1}{2}\omega\tau}{\sin(\frac{1}{2}\omega\tau)} \qquad (9.5)$$

Figure 9.2 shows the SMRR plotted as a function of frequency

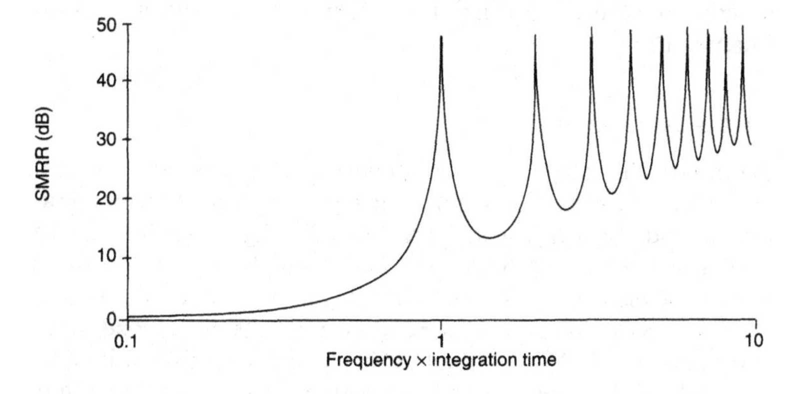

Figure 9.2 *Series mode rejection ratio of an integrating converter.*

and integration period. When $\omega\tau$ is an even and integral multiple of π, then $\sin(\frac{1}{2}\omega\tau) = 0$, and the SMRR is infinite. When $\omega\tau$ is an odd integral multiple of π, then $\sin(\frac{1}{2}\omega\tau) = 1$ and the SMRR is $20\log_{10}(\frac{1}{2}\omega\tau)$. Placing a ruler on Figure 9.2 shows that all the troughs lie on a straight line. That is, on a decibel scale; the minima of the SMRR are proportional to frequency.

In practice, noise rejection may be increased further by adding a low-pass filter before the converter. This is commonly done on the a.c. ranges of a digital voltmeter, but it increases the response time. In high-resolution dvms a filter is essential to obtain a steady reading on all digits, and sophisticated digital filters are used to keep the increase in response time to a minimum.

9.3.2 Improvement of series mode rejection ratio at mains frequency

In the most accurate dvms, the integration period may be locked to the mains frequency. This can be done using a phase-locked loop, which is described in Chapter 4.

9.3.3 Oversampling

An alternative method of measuring the mean voltage is to use a nonintegrating converter, take a number of samples during the integration period, and compute the mean value in the computer. For example, we can obtain a good estimate of the mean value by taking 64 samples, adding them together, shifting the result six places to the right and discarding the six least-significant bits. Indeed, if we do not discard all six LSBs, the quantization error is reduced. (More information on quantization error is given in Chapter 10.)

9.3.4 Systems voltmeters

The top-of-the-range digital multimeters are designed to be used in a complete system, and the user has control over the integration period. Typically, to display 7½ or 8½ digits, the integration period will be a multiple of the mains supply period, but these meters are also able to read at much higher rates at reduced accuracy and resolution – several hundred readings per second at 5½ digits and over 1000 at 4½ digits. Conversion methods which allow this trade-off between speed and accuracy include multiple ramp, charge balance, pulse width and recirculating remainder, which are described in the following sections.

9.4 Dual ramp A/D converter

9.4.1 Principle

Many of the weaknesses of the single ramp design can be eliminated by applying both the input voltage and the reference voltage to the integrator in turn. The simplified circuit for this is shown in Figure 9.3. Each conversion comprises three phases:

1. Integrate from $t = 0$ to $t = \tau_1$. Switch S_1 is closed and the voltage to be converted is applied to the integrator. This phase lasts for a fixed number of clock pulses, usually 1000 or 10 000. This is sometimes called the ramp-up period, but whether the ramp goes up or down depends on the polarity of the input so the

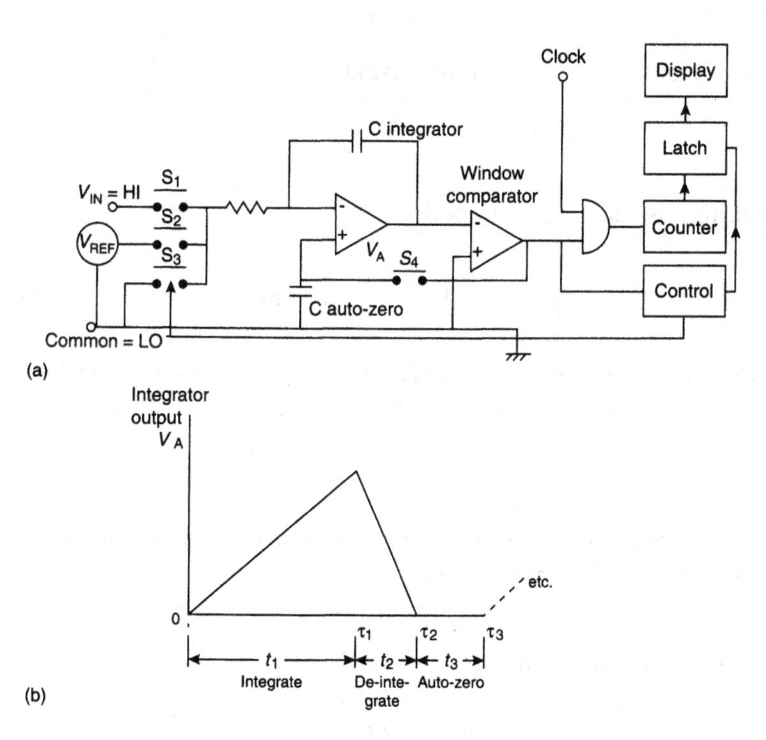

Figure 9.3 *Dual-ramp A/D converter with auto-zero: (a) simplified circuit; (b) waveforms.*

terms integrate and deintegrate are less ambiguous, even though clumsy.

2. The deintegrate period lasts from $t = \tau_1$ to $t = \tau_2$. S_2 is closed so that the reference voltage is applied to the integrator. The polarity of the reference is opposite to that of the input, and this phase lasts until the output of the integrator has returned to zero, as detected by the comparator. During this period the clock pulses are counted and the result is transferred to the latch and display and measures the value of the input.
3. The auto-zero period is from τ_2 to τ_3. S_3 and S_4 are closed and corrections are made for offsets in the circuits, so that the next conversion cycle will start with the integrator output at zero.

Let the integrator output voltage be V_A. During the integration period

$$V_A = \frac{1}{RC}\int V_{IN}\,dt$$

At the end of the integration period $t = \tau_1$

$$V_A = \frac{V_{INav}\,\tau_1}{RC}$$

During the deintegration period

$$V_A = \frac{1}{RC}\left(V_{INav}\,\tau_1 + \int_{\tau_1}^{\tau_2} V_{REF}\,dt\right) \tag{9.6}$$

The end of the deintegration period is determined by $V_A = 0$ which switches the comparator. Therefore,

$$0 = \frac{V_{INav}\,\tau_1 + V_{REF}(\tau_2 - \tau_1)}{RC}$$

Now, from $t = 0$ to $t = \tau_1$ there are n_1 counts (n_1 is a round number) and from $t = \tau_1$ to τ_2 there are n_2 counts. Thus,

$$V_{IN}\,n_1 = -V_{REF}\,n_2$$

and the digital reading is

$$n_2 = -\,n_1\frac{V_{INav}}{V_{REF}} \tag{9.7}$$

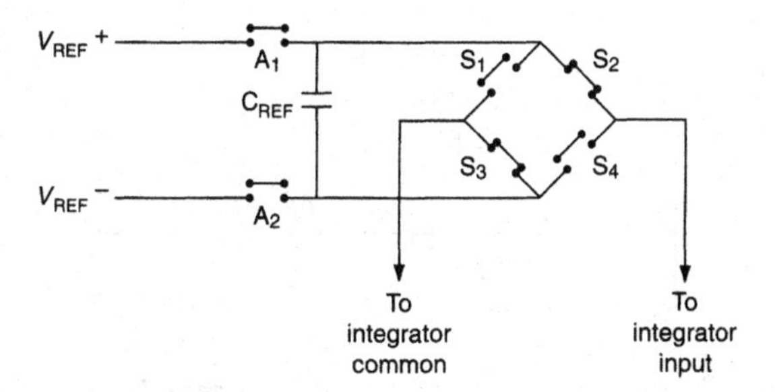

Figure 9.4 *Flying capacitor techniques for auto polarity.*

Note that neither the integrator time constant RC nor the clock frequency occur in this equation, and therefore they are not critical values in the design and do not require high-stability components.

9.4.2 Auto polarity using a flying capacitor

The technique of temporarily storing a voltage on a capacitor can be used to ensure that the reference voltage is always connected in the correct polarity, as shown in Figure 9.4.

During the auto-zero phase, the reference capacitor is charged to the reference voltage through switches A_1 and A_2.

During the integrate phase, the polarity of the input voltage is determined from the polarity of the comparator output, which is the same as the input.

During the deintegrate phase, the reference capacitor is connected to the integrator input through S_1 and S_4, or S_2 and S_3, according to the polarity required.

In a bipolar converter, extra circuitry, which is not shown in Figure 9.3, is included to interface the comparator output levels to the logic circuits. Also, in all but the simplest converters, the low side of the input voltage is isolated from digital ground.

9.4.3 Auto-zero compensation using a flying capacitor

Offsets are unavoidable in op-amps, but their effects can be compensated. This may be done by analogue or digital techniques.

Figure 9.5 shows the flying capacitor technique for auto-zero correction. At the end of each conversion, when the comparator signals the end of the deintegrate phase, the input and reference

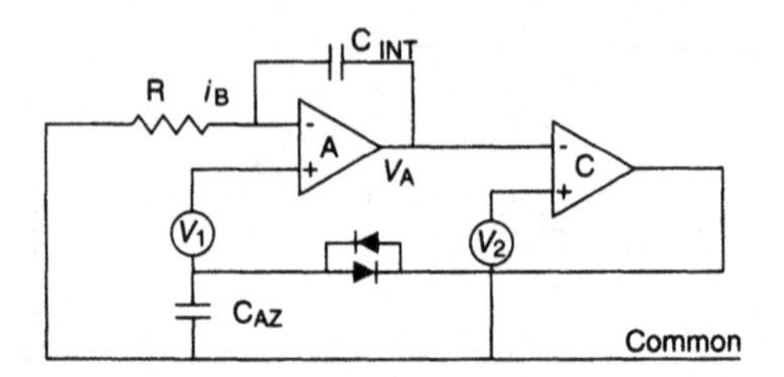

Figure 9.5 *Flying capacitor technique for auto zero: V_1 is the offset due to the integrator (and input buffer if included); V_2 is the offset due to the comparator; i_B is the bias current of the integrator (usually negligible).*

voltages are removed and the input terminal is short circuited to common. Note that the word common is used because this line is not necessarily connected to ground, although it could be.

The effect of the offsets is to swing the comparator output hard positive or negative. Switch S_4 is now closed so that the comparator can drive current into or out of C_{AZ}, which charges up until the voltage on it compensates the offsets. The function of the two back-to-back diodes is to provide increased resistance and hence slower charging as C_{AZ} approaches the correct value. The integrator time constant affects the rate of change of V_A, but not the final value. The charge on C_{AZ} remains throughout the next conversion and is 'topped up' during the next auto-zero phase. The auto-zero capacitor must not be so small that it loses appreciable charge to the stray switch capacitance when S_4 opens. If it is too large, the auto-zero correction is slow, especially when recovering from an overload.

Circuit analysis

The offset in all the input circuits is represented by V_1. This comprises offset in the buffer amplifier, if there is one, offset in the integrator op-amp, thermal emfs in the input circuit with S_3 closed (but not thermal emfs in external circuits), and volt drops due to bias currents flowing through input resistors (these are negligible with FET input transistors).

The voltage V_2 represents the comparator offset, i.e. the voltage which has to be applied at the inverting input in order to bring the comparator output to zero. The circuit will settle towards the condition where V_{AZ} is determined as follows:

$$\text{integrator:} \quad V_A = A_I(V_{AZ} + V_1) \tag{9.8}$$

where A_I is the open loop gain of the integrator.

$$\text{comparator:} \quad V_{AZ} = A_C(V_2 - V_A) \tag{9.9}$$

where A_C is the open loop gain of the comparator.

Substituting (9.8) in (9.9) gives

$$V_{AZ} = A_C(V_2 - A_I(V_{AZ} + V_1))$$

$$V_{AZ}(1 + A_C A_I) = A_C(V_2 - A_I V_1)$$

since both A_C and A_I are much greater than unity,

$$V_{AZ} = \frac{V_2 - A_I V_1}{A_I}$$

$$= \frac{V_2}{A_I} - V_1 \tag{9.10}$$

Flying capacitor auto zero is incorporated in several integrated circuit designs of dual-ramp converter, including Intersil 7106 and 7107.

Digital auto zero is used in voltmeters which incorporate a microprocessor. The input terminals are short-circuited and a reading is taken. This is stored and subtracted from subsequent readings. Example 9.5 illustrates this.

9.4.4 Effect of delays in the switching and in the comparator

At the end of the integrate period, S_1 opens and S_2 closes (Figure 9.3). Obviously, S_2 must not close too soon and connect the reference voltage to the input, and some designs insert a short delay between the two phases and allow for it in the counting.

At the end of the deintegrate period the operation time of the comparator is important. Its output voltage should be

$$V_C = -V_A A_C \tag{9.11}$$

where V_A is the output voltage of the integrator and $-A_C$ is the gain of the comparator – the negative sign implies an inverting comparator.

The comparator output will actually be slightly delayed as illustrated in Figure 9.6(a), which shows the end of the deintegrate period in detail. The solid line V_A represents the output of the

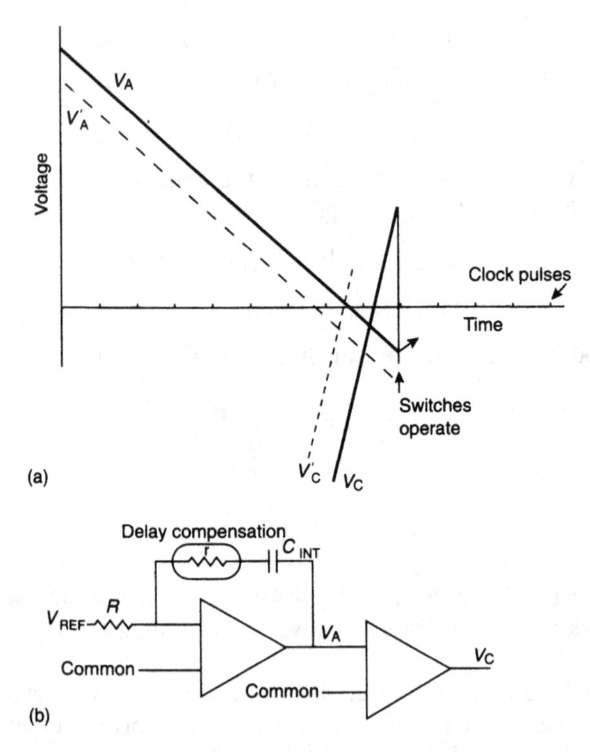

Figure 9.6 *Effect of delay in the comparator: (a) voltages, V_A = integrator output and V_C = comparator output; (b) correction circuit.*

integrator returning towards zero, and the solid line V_C shows the comparator output voltage rising more steeply, but later. In the example a clock pulse occurred after V_A had crossed zero but before V_C reached zero, so the switches did not operate until the next clock pulse, making the resulting count one too large. Thus this delay places an upper limit on the clock frequency and on the resolution of the converter. For example, a 4½-digit voltmeter with a deintegration time of 40 ms requires a clock frequency $> 20\,000/0.04 = 500$ kHz, and the comparator delay must not exceed 2 μs (which is not difficult).

Delay in the comparator may be partially compensated by adding a small resistor r in series with the integrating capacitor C_{INT}, as shown in Figure 9.6(b). The effect of this is to reduce V_A by $V_{REF}r/R$ where R is the integrator input resistance. The rate of change of V_A is V_{REF}/RC_{int}. Therefore, V_A' and V_C' each cross zero earlier, as shown by the dashed lines. The change in zero-crossing time is given by

$$\frac{V_{REF}\, r/R}{V_{REF}/(RC_{INT})} = rC_{INT} \tag{9.12}$$

In practice, r will probably have to be found by experiment, but is likely to be of the order of 10 Ω.

9.4.5 *Dielectric absorption*

The capacitance C is given by the formula

$$C = \frac{\mu_0 \mu_R A}{d} \tag{9.13}$$

The term μ_0 is a fundamental constant known as the 'permeability of free space' $= 4\pi \times 10^{-7}$ H m^{-1}, A is the area of, and d is the distance between, the plates. The term μ_R, the relative permeability, is a parameter of the dielectric material and is often taken as a constant but, in fact, it varies with frequency. To understand why, we must look more closely at what happens when a capacitor is charged.

When a voltage is applied to the plates of a capacitor, it establishes an electric field across the dielectric. This field polarizes the dielectric; that is, it causes any charged bodies within the dielectric to move and any molecules which are not electrically symmetrical to twist and try to align themselves with the field. In Figure 9.7, C_1 is the main capacitance and R_1 is the insulation resistance.

The four polarization mechanisms are:

- **Electronic** – this is the displacement of electrons. It happens in times of the order of the period of light and determines which wavelengths are absorbed and which are transmitted.

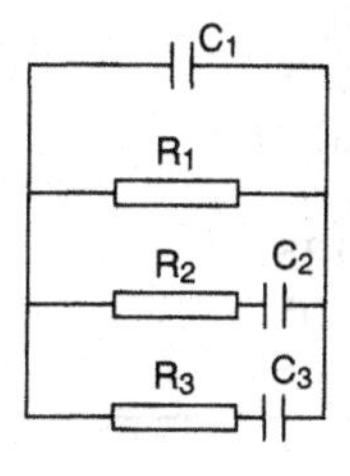

Figure 9.7 *Imperfections in a capacitor.*

- **Atomic** – this refers to the relative displacement of atoms in a compound.
- **Dipole** or **orientational** – molecules which have an asymmetric charge distribution and hence a permanent dipole moment try to align themselves with the applied field. Such materials are described as polar. R_2C_2 is the time constant of capacitance due to orientational polarization.
- **Space charge** or **interfacial** – if there are any free ions in the material they will drift in the direction of the field. R_3C_3 is the time constant of the capacitance due to space charge polarization.

All polarization takes time – from electronic polarization which occurs in femtoseconds (10^{-15} s) to space charge polarization which may take hours. The effect of delay in polarization may be represented by the circuit of Figure 9.7.

Analysis of the above equivalent circuits with an alternating applied voltage shows that energy is absorbed when the period of the electric field is equal to any of the time constants, and that the magnitude of the total capacitance varies with frequency.

The effect of delayed polarization is also known as **dielectric absorption** because, if a capacitor is charged and then discharged, the charge recovered is less than the charge supplied (as if some of the charge had been absorbed by the dielectric in the way that a sponge absorbs water). It is obvious that the accuracy of a dual-ramp A/D relies on all the charge being recovered during the deintegrate phase, so the integration capacitor must always be made of a dielectric which does not suffer from absorption.

Table 9.1 *Dielectric absorption*

Material	*Dielectric absorption (%)*	*Insulation resistance (MΩ)*
NPO ceramic	0.6	10^5
Stable ceramic	0.25	10^6
Mica	0.3–0.7	10^2
Polyester	0.5	10^4
Polycarbonate	0.35	10^5
Polypropylene	0.05	10^5
Polystyrene	0.05	10^6

The insulation resistance and dielectric absorption of all the common dielectrics are shown in Table 9.1. Theses figures, which are taken from the more comprehensive table of dielectric parameters in Appendix C, show that the best dielectrics for integrators are polythene and polypropylene. Polyester and polycarbonate are polar materials.

The theory of dielectrics is discussed in more detail in Daniel (1967), Kaye (1986) and von Hippel (1994).

9.5 Triple ramp and multislope A/D converters

The triple ramp converter is a modification of the dual ramp, designed to reduce errors introduced by comparator delay and by dielectric absorption. The integration phase is the same as in the dual ramp, and lasts for time t_1. The deintegrate phase is subdivided into two sections:

- In the first section, reference voltage 1 is applied. This is larger than would be used in a dual ramp so charge is removed from the capacitor as rapidly as possible, thus reducing errors due to dielectric absorption. This lasts for time t_2 until V_A reaches V_{REF}/k.
- When this level is detected, the reference voltage is reduced from V_{REF1} to $V_{REF2} = V_{REF1}/k$, and the frequency of clock pulses fed to the counter is reduced in the same ratio. In a 12-bit binary counter, the changeover could occur after 6 bits which would make $k = 64$, or after 8 bits which would make $k = 256$. When the comparator finally detects the zero crossing, V_A is changing slowly and the zero-crossing time can be accurately determined. The time for the second half of the deintegrate phase is t_3.

The equation describing the operation is

$$\frac{V_{INav} t_1}{RC} + \frac{V_{REF} t_2}{RC} + \frac{V_{REF} t_3}{kRC} = 0 \qquad (9.14)$$

in which the term RC cancels throughout. Now,

$t_1 = n_1/f_C$, which is fixed

$t_2 = n_2/f_C$

$t_3 = n_3 k/f_C$

Substituting gives

$$\frac{V_{\text{INav}}n_1}{f_C} + \frac{V_{\text{REF}}n_2}{f_C} + \frac{V_{\text{REF}}(n_3 k/f_C)}{k} = 0$$

which simplifies to

$$V_{\text{INav}}n_1 + V_{\text{REF}}n_2 + V_{\text{REF}}n_3 = 0$$

$$V_{\text{INav}} = -V_{\text{REF}}\left(\frac{n_2 + n_3}{n_1}\right) \tag{9.15}$$

Thus, as in the dual ramp, the count is independent of the integrator time constant and the clock frequency. It is important that the change of reference voltage and the change of counter input frequency occur simultaneously.

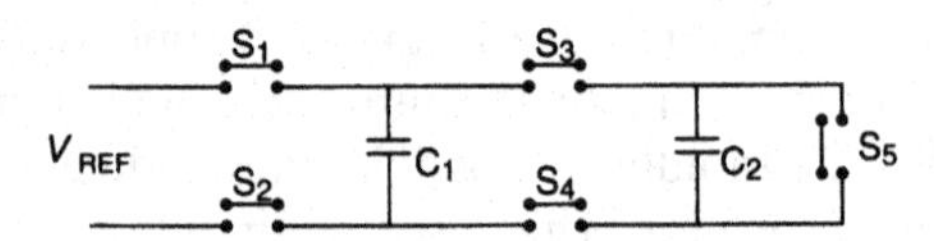

Figure 9.8 *Triple ramp converter reference voltage.*

The flying capacitor technique may be used to reduce the reference voltage in a known ratio, as illustrated in Figure 9.8. The sequence is:

1. During t_1 (integration), S_1 and S_2 are closed and C_1 is charged to V_{REF}. S_3 and S_4 are open and S_5 closed so that there is no charge on C_2.
2. During t_2 (initial deintegration), S_1, S_2 and S_5 are open.
3. During t_3 (final deintegration), S_3 and S_4 are closed and charge is shared between C_1 and C_2.

Capacitors can be formed in integrated circuits with a ratio which is known much more accurately than the actual value of either capacitance.

The triple ramp converter is not widely used, but the principle has been extended to the **multislope converter** which is used in very high-accuracy digital voltmeters made by Wavetek-Datron and by Hewlett-Packard. Both instruments achieve 8½-digits decimal resolution, which is equivalent to 28-bit binary. Also, the end of the first deintegrate ramp is detected at zero. The ramp overshoots and the second reference voltage is applied with the opposite polarity to the first to bring V_A back towards zero. This is followed by a third ramp in the opposite direction again. Hewlett-Packard use $k = 10$ (Stever, 1989 and Goeke, 1989); Wavetek-Datron use $k = 16$ (Chenhall, 1987). To achieve the highest possible accuracy is not as simple as the above description implies, so such companies incorporate a number of other refinements to their designs. High accuracy demands not only high resolution, but linearity and reference voltage stability of the same order.

Multislope converters have the further advantage that an approximate result can be obtained very quickly and so the trade-off between speed and resolution is under the control of the user, according to the number of ramps used.

An integrated version of the multislope converter is the AD 1175K from Analog Devices with a resolution of 22 bits.

9.6 Voltage to frequency and charge balance A/D converters

The dual ramp is actually one form of charge balance converter because the charge put into C_{INT} during the integrate phase is equal to the charge removed during deintegrate, but the name 'charge balance' is more usually applied to techniques which alternate charge and discharge phases and thus reduce the effect of dielectric absorption.

9.6.1 Bipolar charge balance converters

The converters described below are unipolar; that is, the input voltage must always be of the same polarity. To make a charge balance converter accept either polarity, an additional, constant reference current is fed to the summing junction such that zero frequency corresponds to maximum negative input, maximum frequency corresponds to maximum positive input, and half maximum frequency corresponds to zero input.

9.6.2 Voltage to frequency methods

Voltage to frequency converters generate a train of pulses at a rate proportional to V_{IN}. If these are counted for a fixed time, the count is a measure of the mean value of V_{IN} during that time. The pulse train can be transmitted over a long distance without loss of accuracy, and analogue and digital grounds can be completely isolated.

9.6.3 Simple voltage to frequency converter

The simplest voltage to frequency type is a derivative of the single ramp design but with V_{IN} and V_{REF} interchanged. This is illustrated in Figure 9.9(a). The integrator output rises at a rate proportional to V_{IN} and, when it reaches V_{REF}, the comparator fires the monostable and discharges C_{INT}. This type is not very accurate because C_{INT} discharges exponentially and is therefore not completely discharged within the short duration of the monostable pulse. Each pulse discharges $C_{INT}V_{REF}$ coulombs, so the balance equation is approximately

$$fC_{INT}V_{REF} = \frac{V_{IN}}{R} \tag{9.16}$$

Allowing for the discharge time of the monostable t_m,

$$f = \left(RC_{INT}\frac{V_{REF}}{V_{IN}} + t_m\right)^{-1} \tag{9.17}$$

9.6.4 Charge balance voltage to frequency converter

The charge-balance type shown in Figure 9.9(b) is an improvement on Figure 9.9(a) because each time the comparator operates the reference current flows into the integrator for a precisely determined time. In other words, it delivers a precise charge to the integrator. The balance equation of this type is derived as: current into summing junction equals current out of junction:

$$\frac{V_{IN}}{R} = i_{REF}\,\delta t f$$

where δt is duration of charge pulse and f = frequency of pulses, to give

$$f = \frac{V_{IN}}{R\,i_{REF}\,\delta t} \tag{9.18}$$

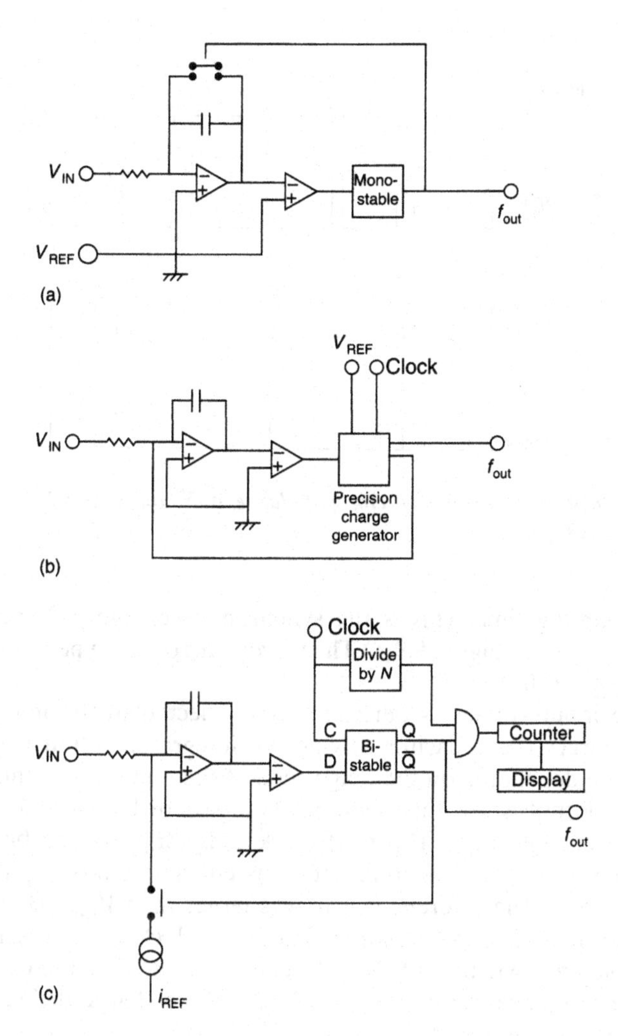

Figure 9.9 *Voltage to frequency converter circuits: (a) ramp type; (b) charge balance; (c) synchronous charge balance.*

An example of the charge-balance type in integrated form is the AD 650 from Analog Devices.

9.6.5 Synchronous charge balance voltage to frequency converter

A further improvement in linearity, and hence accuracy, may be achieved by synchronizing the duration of the discharge pulse to

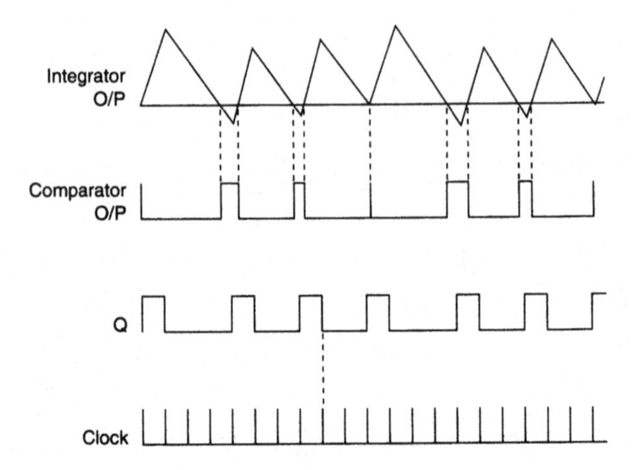

Figure 9.10 *Waveforms in a charge balance Δ–Σ converter, when* $V_{IN}/R = 0.3\, i_{REF}$.

the counting time. This is the synchronous charge balance converter shown in Figure 9.9(c). The waveforms of this type are shown in Figure 9.10.

The input voltage is permanently connected to the integrator and charges the capacitor driving the integrator output negative at a rate depending on the magnitude of the input. When the integrator output is less than zero, the comparator holds the D input of the bistable high. The next clock pulse triggers the bistable, which switches on the reference current and sends a pulse to the counter. The reference current is larger than V_{INmax}/R, so the capacitor begins to discharge. The next clock pulse resets the bistable and switches off the reference current. Thus the duration of the reference current pulse is $1/f_c$, and the charge delivered to the capacitor is i_{REF}/f_c.

When V_{IN} has replaced the charge removed by the reference current the comparator triggers the bistable again and the process is repeated. This may be summarized by the charge balance equation, which is

$$f = \frac{V_{\mathrm{IN}} f_c}{R\, i_{\mathrm{REF}}} \tag{9.19}$$

This part of the circuit is called a delta modulator. The counting time is N clock periods so the count is given by

$$n = \frac{V_{\mathrm{INav}} N}{i_{\mathrm{REF}} R} \tag{9.20}$$

which is independent of clock frequency and integration capacitance. A further advantage of the charge balance types over the dual ramp is that V_{IN} is measured continuously and not intermittently. It is, of course, possible to choose the gate time, which is also the integration time, as a multiple of the mains period to improve the SMRR at mains frequency. Finally, the gate time may be used as a range adjustment. Thus if N is increased by a factor of 10, the full-range input voltage is reduced by a factor of 10 also. An integrated version of this converter is AD652, made by Analog Devices.

This design is also known as delta–sigma (Δ–Σ) (Owens, 1982). The circuit generating the pulse train is a delta modulator and the counter finds the sum or Σ of the pulses. Some people prefer Σ–Δ on the grounds that 'sum of the bits' makes more sense than 'bits summed'.

9.6.6 Audio sigma–delta A/D converters

The development of digital audio systems has led to a demand for high-resolution converters with a sampling rate of more than 40 000 samples per second, such as 44.1 kSPS for compact discs and 48 kSPS for digital audio tape. A few years ago, successive approximation would have been the obvious choice, but differential nonlinearity at the MSB transition can cause problems. The sigma–delta converter as described above is a slow but accurate converter for instrumentation applications and can do single conversions. It has been further developed for hi-fi audio systems requiring a bandwidth of 20 kHz and continuously repeated conversions. A delta modulator actually samples the input at half the clock frequency and generates a single-bit pulse train as illustrated in Figure 9.10. For audio converters, the pulse train is applied to a digital low-pass filter instead of being counted for a fixed integration time. This filter has a high resolution of 16 or 20 bits, and is known as a decimation filter because its output is read at the lower sampling rate of 48 kSPS or thereabouts. Typical parameters are clock frequency 6.144 MHz, sampling frequency 3.072 MHz, output sample rate 48 kSPS and resolution of 16 bits. This is described as 64 times oversampling ($64 \times 48 = 3072$). As these designs are specifically intended for digital audio circuits, they are made in

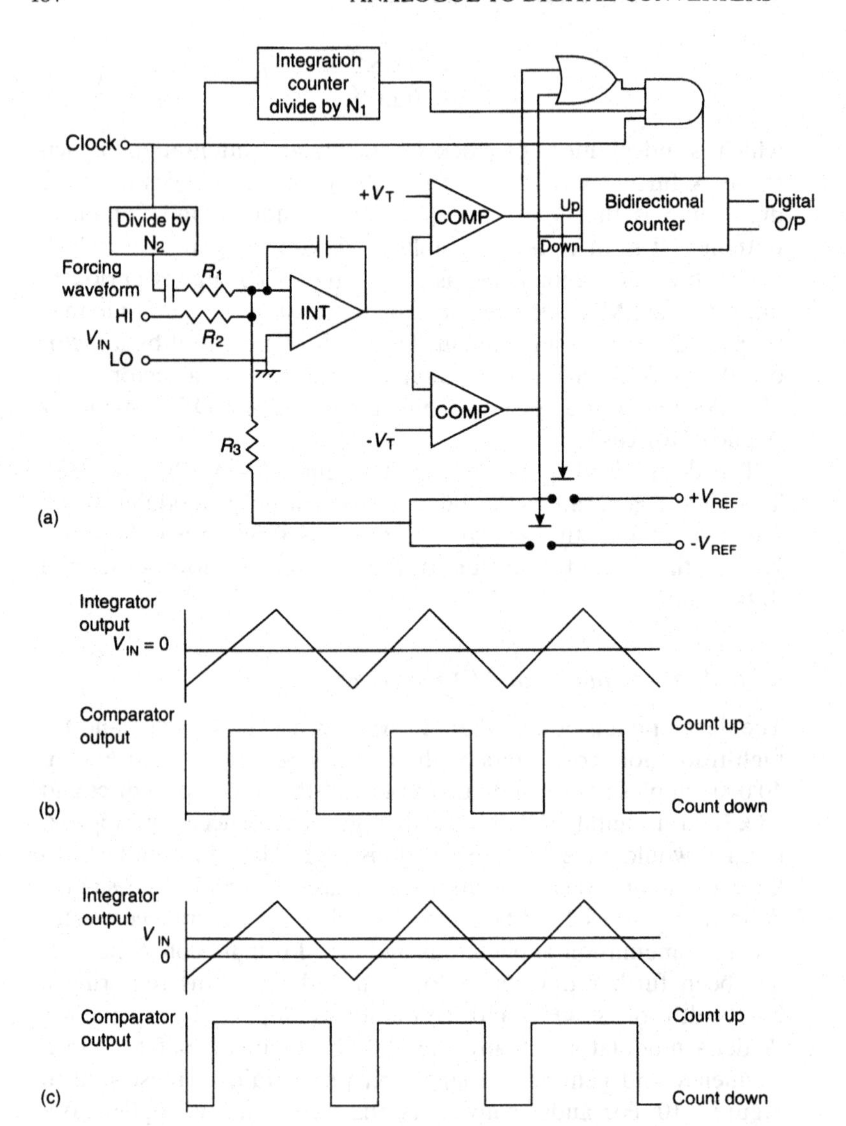

Figure 9.11 *Pulse width converter: (a) simplified circuit; (b) waveforms when $V_{IN} = 0$; (c) waveforms when V_{IN} is positive.*

pairs for the two stereo channels, and examples of this type of converter are – Crystal CS 5326, Burr-Brown PCM1760/DF1760 and Analog Devices AD 1848. For background information see Mar and Regimbal (1990), Dijkmans and Naus (1987), Analog Devices (1989–90), Glover *et al.* (1991) and Johnston (1991).

9.7 Pulse-width A/D converters

Figure 9.11 shows the simplified circuit diagram and waveforms of the pulse width converter which is favoured by Solartron for their high-resolution digital voltmeters (Pearce, 1983; 1987); UK patents 1273790 and 1434414). This technique can also achieve 8½-digit resolution.

When the input voltage is zero, the output of the integrator is a triangular wave which is symmetrical about zero, and both comparators are ON for equal times. The counter counts UP and DOWN equal numbers and, on balance, remains at zero. The integration period lasts for many periods of the forcing voltage. When an input voltage is applied, the symmetry is upset. The final count may be calculated by considering the charge balance at the integrator input. Let:

t_1 be the time the positive reference is ON in each forcing period
t_2 be the time the negative reference is ON in each forcing period
τ be the period of the forcing voltage
N_1 be the number of clock pulses in the total integration time
N_2 be the number of clock pulses in each forcing period

The charge balance equation is

$$\frac{V_{\mathrm{REF}}t_1}{R_3} - \frac{V_{\mathrm{REF}}t_2}{R_3} + \frac{V_{\mathrm{IN}}\tau}{R_2} = 0 \qquad (9.21)$$

$$\frac{t_2 - t_1}{\tau} = \frac{V_{\mathrm{IN}}R_3}{V_{\mathrm{REF}}R_2}$$

Now, $\tau = N_2/f_c$; therefore, the number of clock pulses n_f counted in each forcing period is

$$n_f = (t_2 - t_1) \times f_c$$

As in the sigma–delta circuit, a bistable is used to ensure that n_f is an integer. The total count is the number of pulses in N_1/N_2 forcing periods n_t:

$$n_t = \frac{N_1 V_{IN} R_3}{V_{REF} R_2}$$

When $R_2 = R_3$, this simplifies to

$$n_t = \frac{N_I V_{IN}}{V_{REF}} \tag{9.22}$$

9.8 Pulse-width, pulse-height wattmeter

A circuit similar to the one just described generates a pulse train such that

$$\frac{t_2 - t_I}{t_I + t_2}$$

is proportional to the instantaneous load voltage v. This is used to switch a voltage proportional to the instantaneous load current i. The result is an analogue output whose mean value is proportional to the instantaneous power, vi. The switching frequency is removed by a low-pass filter to give an analogue voltage proportional to mean power. As both the load voltage and current are varying at 50 Hz, the frequency of the forcing signal must be very much higher, i.e. in the kilohertz range.

9.9 Successive approximation A/D converters

9.9.1 Principle

Consider again the counter type converter described on page 163. Its major disadvantage is its slowness, because it counts from zero for each conversion. Like the counter, the successive approximation converter generates a digital code and compares the output of a D/A to the input voltage. Conversion times are of the order of microseconds, so this type of converter is suitable for the majority of data acquisition applications, provided that it is preceded by a sample and hold circuit.

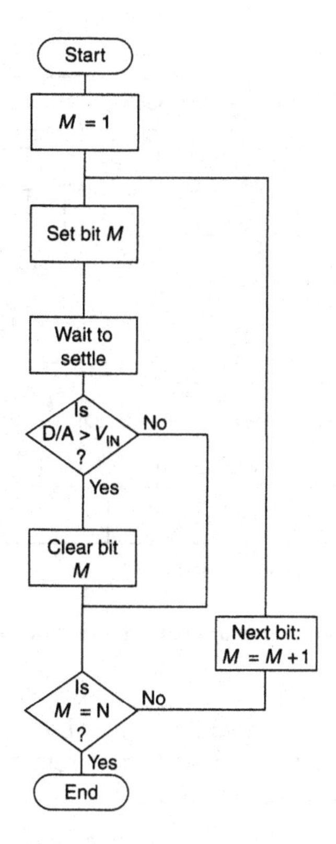

Figure 9.12 *Successive approximation flowchart (M = number of the current bit, MSB = 1, N = total number of bits).*

The operation of a successive approximation converter follows the flowchart shown in Figure 9.12. This can be implemented in a computer or in hardware.

9.9.2 Resistive ladder converters

A suitable circuit is shown in Figure 9.13. Integrated versions are available from all the major manufacturers up to 12-bit resolution, and up to 18-bit from some of them. The first clock pulse (T_0) sets the MSB which makes the output of the D/A = $\frac{1}{2}V_{REF}$. When this has settled, it is compared with the input, and if the $V_{D/A} > V_{IN}$, the MSB is erased on the next clock pulse (T_1). If $V_{D/A} < V_{IN}$, the MSB is retained.

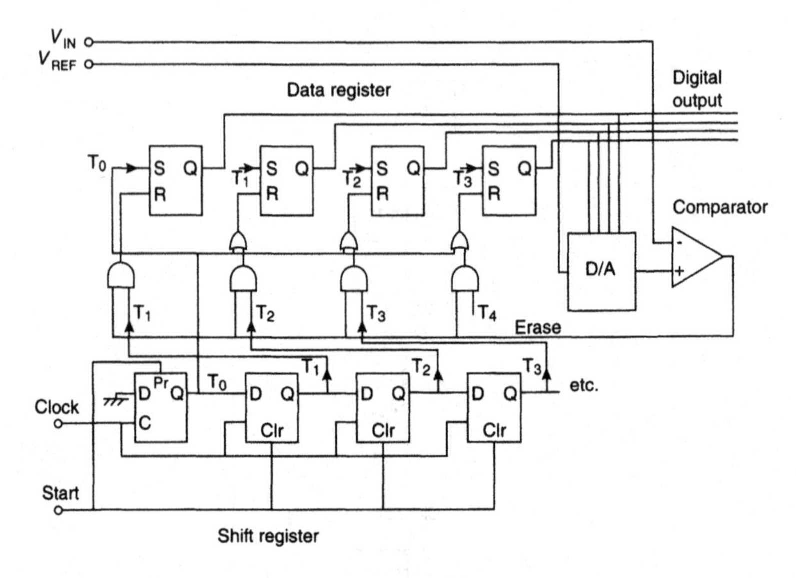

Figure 9.13 *Successive approximation converter circuit.*

The T_1 pulse also sets the next bit. The process continues until all the bits have been tried and retained or erased as required. The conversion time is thus:

$$t_c = \frac{N+1}{f_c} \tag{9.23}$$

The upper limit to the clock frequency depends on the settling time of the D/A and the speed of the comparator. It is usually of the order of several hundred kilohertz and the conversion time is typically in the range 2–100 μs.

9.9.3 Sampling successive approximation converters

Any A/D which has a built-in sample and hold may be classed as a sampling converter, but this name is particularly applied to a type of successive approximation converter capacitor in place of resistors in the divider. In one design (Johnston, 1991), a flying capacitor first captures the input voltage in a similar way to a sample and hold circuit (Chapter 5). When conversion starts, it shares the charge with other capacitors, following a successive approximation sequence. The use of a capacitive voltage divider instead of the resistance ladder is claimed to improve the temperature coefficients of the offset and gain errors. Examples are Crystal CS5012, Burr-Brown ADS774 and Analog Devices AD7882.

9.10 Recirculating remainder A/D converters

Another technique of adjusting the output of a D/A converter until it is equal to the input is the recirculating remainder patented by Fluke Ltd. It was designed for digital voltmeters and works in a decimal instead of a binary base. The comparator in the successive approximation design is replaced by a differential amplifier which compares the output of the D/A with the input and multiplies it by 10. The conversion process is as follows.

The MSD of the D/A is incremented until the difference is less than one digit. At this stage, the amplifier output which is 10 times the difference is known as the remainder. The values of both the MSD and the remainder are stored. The D/A is then reset and the process repeated, comparing the remainder with the D/A output to find the second digit and a new remainder. The comparison is repeated until all the values of all the digits are known. This technique has been used for digital voltmeters of at least 6½-digit resolution. It has the same advantage as the multislope technique that an approximate result is obtained very quickly, and the user can control the trade-off between speed and accuracy.

9.11 High-speed A/D converters

9.11.1 *Flash converters*

For conversion rates greater than 10^6 samples per second (MSPS), special converters and special techniques are required. Emitter coupled logic is used, and great care has to be taken with ground planes (Pratt, 1974; Smith, 1974). The advent of digital image processing which includes synchronization, slow motion, standards conversion and special effects, has created a big demand for converters that can digitize colour television pictures. The conversion rate needed can easily be estimated:

25 frames per second, 625 lines per frame,
$625 \times 4/3$ pixels per line = 13 MSPS

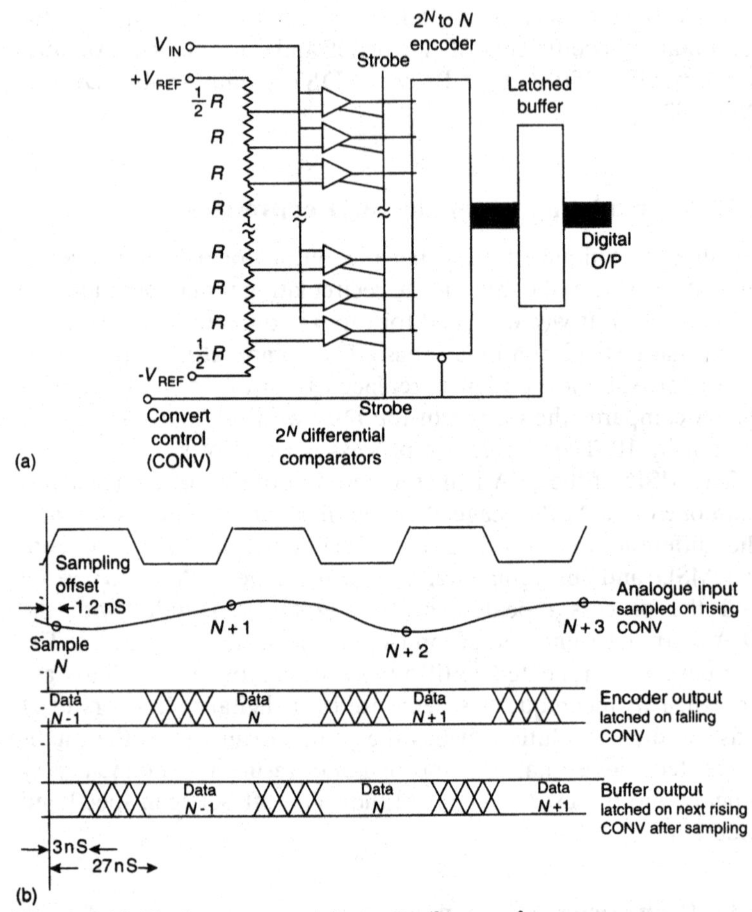

Figure 9.14 *Flash converter: (a) circuit; (b) waveforms.*

In the PAL system used in the UK, the luminance signal is actually sampled at 13.5 MSPS, and the blue–yellow and red–yellow colour difference signals are sampled at 6.75 MSPS each, making a total sampling rate of 27 MSPS. An 8-bit A/D converter brings the bit rate to 216 Mbps (Slater, 1994). A board designed to digitize signals from a video camera or recorder, store one frame in a buffer, and send it to the PC bus is known as a **frame grabber**.

For the highest possible speed a flash or parallel converter as shown in Figure 9.14 is used. A chain of resistors connected across the reference voltage provides taps at the transition voltages between successive levels. Thus the lowest tap in the chain is ½ LSB and the next tap is 1½ LSB above the $-V_{REF}$ line. Any input between these two levels is digitized as 1 LSB. The input voltage is compared with all the taps simultaneously by the comparators. All the comparators for levels below the input voltage are ON and all the comparators for levels above the input are OFF. The encoding logic, which is emitter-coupled for speed, converts the comparator outputs into binary code. For example, the Boolean equations (overbar indicates NOT) for a 3-bit converter are

$$\text{MSB} = 4$$

the MSB is ON if comparator 4 is ON

$$\text{middle bit} = 6 + \bar{4}.2$$

The middle bit is ON if comparator 6 is ON, *or* if comparator 2 is ON *and* 4 is OFF.

$$\text{LSB} = 7 + \bar{6}.5 + \bar{4}.3 + \bar{2}.1$$

The LSB is ON if the highest comparator which is ON is odd-numbered.

The logic becomes rapidly more complex as the number of bits increases: the number of comparators is $2^N - 1$ for N bits; logic for 3 bits requires $2(2^N - 1 - N) = 8$ gates; 4 bits require 22 gates; 8 bits require 494 gates; and 12 bits require 8166 gates. Although these numbers may be reduced by using wired-OR connections, it remains true that the complexity increases very rapidly with word length.

It is obvious that the digital code is not valid until all the comparators have settled. Usually the comparator outputs are read into the encoder a short time after the 'convert' command and the digital code is transferred to the output latches after another short delay. A short delay in this context is a matter of nanoseconds. The timing for a series of samples in rapid succession is shown in Figure 9.14(b). The rising edge of clock pulse 1 strobes the comparators and the falling edge strobes the encoder. The rising edge of pulse 2 latches data to the output buffer. The data is available after a further short delay (typically 27 ns). This technique is known as **pipelining** and, although each conversion requires two clock pulses, new data appears at the output every clock pulse. The specification of this type of converter includes a figure for sampling rate instead of conversion time.

For a flash converter, **bandwidth** is defined as the maximum frequency, full-range input sinewave that can be accurately digitized.

Conversion rates of 125 MSPS are claimed for the AD 9002. Specialized A/Ds for fast sampling oscilloscopes were reported to have achieved sampling rates of 1 GHz as long ago as 1987 (Locke, 1987).

9.11.2 Two-stage flash A/D converters

The hybrid, sub-ranging or two-stage converter in Figure 9.15 gives increased resolution without such a great increase in logic complexity, while retaining almost the speed of a true flash converter. For example, 8-bit resolution can be obtained by combining two 4-bit flash converters.

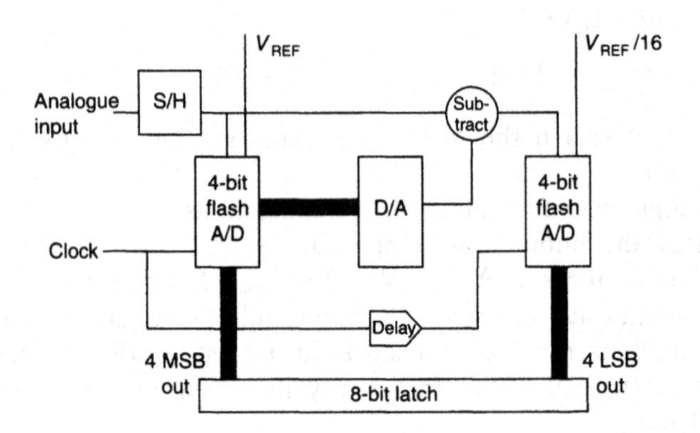

Figure 9.15 *Two-stage flash converter.*

Basically, the operation is as follows. The four most significant bits are encoded and the result clocked into the most significant half of the data register. These bits are reconverted by a high-speed D/A converter. This voltage is subtracted from the input and the difference is encoded by a second flash converter and clocked into the least significant half of the data register. It is necessary to include a sample and hold circuit in the analogue input to the subtractor and a delay in the clock signal to the LSB to compensate for the small but unavoidable delay in the first encoder and the D/A.

For this scheme to work, all the components have to be accurate to 8-bit accuracy. This is a stringent requirement and such a converter would be likely to have nonlinearities each time the MS byte changed by one bit. Digital subranging is a technique to compensate for errors at the changeover points between the two halves of the output register by using an extra bit. Suppose the MS A/D has a small error equivalent to one-eighth of its own LSB, which is equal to 2 LSB of the LS converter. The LS byte can have any value from 0 to 15, so the output of the summer can vary from –2 to + 17 bits of the LS A/D, whereas the range of a 4-bit A/D is only 0 to 15 LSB. Therefore an extra bit is required for the LS A/D. So an 8-bit converter is made up of 4 MSB and 5 LSB and the results are combined logically. Similarly, a 12-bit converter is made up of 6 MS + 7 LS bits. Scheingold (1986) explains the technique thus:

> The result of the first conversion – in the holding register – is equal (digitally) to: (Input – Residue + 6-bit error). It is converted to analog and subtracted from Input. Assuming a perfectly accurate DAC and subtraction, the result is (Residue – 6-bit error). That is, $I - (I - R + E6) = R - E6$. This difference is scaled up and converted, giving a digital quantity equal to (Residue – 6-bit error + 12-bit error). Finally, the two digital words are added, giving: Input + 12-bit error. That is,
>
> $$(I - R + E6) + (R - E6 + E12) = I + E12$$

The principle has been extended to three stages: two 5-bit stages combine to give 9-bit resolution and a final 4-bit stage gives 12 bits overall. This circuit (AD9023) can convert at 20 MSPS.

9.12 Interfacing a converter to a microprocessor

9.12.1 Data transfer methods

Converters which may be connected directly to the bus lines are described as **microprocessor compatible**. Noncompatible converters may also be used with a microprocessor, but they need to be connected via an interface adapter. More details about computer busses and interface adapters are given in Chapter 11.

In each case, the program must send a 'start conversion' command to the A/D, wait until conversion is complete and then read the data. There are several methods of determining when conversion is complete. The most primitive is to make the computer wait for a period longer than the conversion time. Apart from that there are four methods of data transfer:

1. *Programmed input/output with software polling* Having started the conversion, the program goes into a loop effectively saying to the A/D 'Have you finished yet?' until it gets a 'Yes' answer, then it transfers the result of the conversion, either one byte or two, into the processor.
2. *Progammed input/output with interrupt* The 'conversion complete' pin on the A/D is connected to the interrupt pin on the microcomputer, and the program arranged so that when the interrupt occurs, the program jumps to a subroutine to read the data. This method is particularly suitable for slow (e.g. integrating) converters because the program can do other tasks such as gain or offset correction or linearizing while the next conversion is in progress.
3. *Direct memory access (DMA)* Special hardware is needed, but this method is faster because data is transferred from the data bus to the memory without passing through the processor, and up to 32 000 readings can be transferred in one burst.
4. *Double buffered input* For even higher rates the converter may pass its data into one of two buffers on the converter circuit board, which operate alternately. One fills up with data from the A/D while the other transfers the previous batch into the computer memory. Then the buffers change over and the converter continues without interruption. The maximum data transfer rate depends on the number of bytes to be transferred per reading and on the clock frequency of the processor. For comparison, the sampling rates below are typical and are quoted by Data Translation (1993):

programmed I/O with software polling	20 kHz
programmed I/O with interrupt	40 kHz
single channel DMA	100 kHz
dual channel DMA	250 kHz

9.12.2 Mircoprocessor-compatible converters

The data output lines on a compatible converter can be connected directly to the data lines of the microprocessor. To make this possible all the data lines use three-state logic. A typical arrangement is shown in Figure 9.16, based on a 12-bit AD574 connected

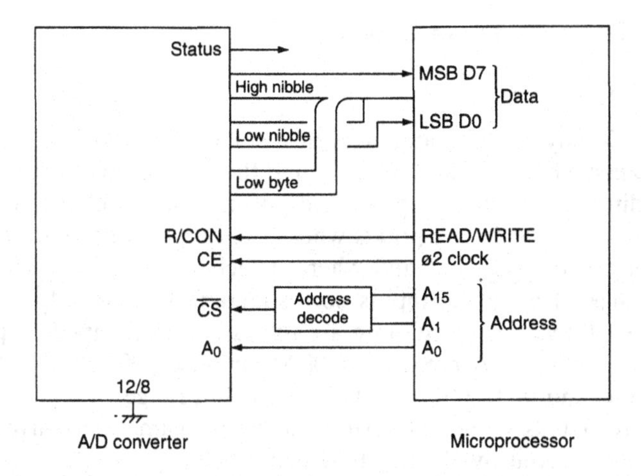

Figure 9.16 *Interfacing an A/D to a computer.*

to an 8-bit 6800 series microcomputer. The 12/8 pin is connected to ground to enable 12-bit conversions to be read into two 8-bit locations.

Microprocessor-compatible A/Ds have a 'chip select' ($\overline{CS}$) pin. When this is held low by the address decoder the chip is active, otherwise all the data lines are in the high impedance state and do not affect the operation of the bus. The converter appears to the microcomputer as if it were one or more memory locations. In this example, the A/D is allocated two addresses: the even-numbered address controls the high byte and the odd-numbered one controls the low byte. When the microcomputer is programmed to read from or write to the selected address, the address lines select the chip (driving $\overline{CS}$ low), the READ/WRITE line goes low for write taking with it the READ/CONVERT pin of the A/D. Finally the phase-2 clock pulse, $\phi 2$, connected to the chip enable (CE) of the A/D, starts the conversion. The computer then waits until conversion is complete, either by a delay loop or it reads the status line of A/D. Then it reads the data by reading the two addresses as if it were reading data out of memory. As eight bits comprise a byte, four bits are known as a **nibble**. The high byte is read into an even-numbered address (A0 = 0) and the high nibble of the low byte with four spare zeros at the LSB end is read into the next address (A0 = 1).

There are many variations of this basic idea and detailed diagrams are given in the A/D manufacturers' data sheets.

9.13 Testing A/D converters

9.13.1 Static tests

An A/D may be calibrated against a standardized potentiometer or against a better quality A/D. To find the analogue voltage corresponding to a particular digital output, measurements should be made at the two **flicker points** where the output flickers from the test value to 1 LSB less and where it flickers to 1 LSB more, i.e. the range of analogue inputs corresponding to that output. The most useful values to calibrate are the same as described on page 152 for testing D/A converters: 0000 for offset; 0001, 0010, 0100, 1000 etc. and 0001, 0011, 0111, 1111 for full range.

More details of tests for successive approximation converters have been given by Young (1987 and 1990).

9.13.2 Offset error

Measure the flicker point at 0–1 LSB. This should be at ½ LSB. Subtract ½ LSB to find the offset. Note that in a unipolar converter 0 LSB = 0 V, but in an offset binary converter 0 LSB is full range voltage negative. A digital voltmeter presents its result in terms of sign + magnitude and so the offset is measured at 0 V.

9.13.3 Gain error

Measure the flicker points at the ends of the range, i.e. between 0 LSB and 1 LSB and between full range (V_{REF} – 1 LSB) and overload. The difference should be (V_{REF} – 1 LSB):

$$\text{gain error} = \frac{\text{measured range} - \text{true range}}{\text{true range}} \times 100\%$$

$$= \left[\frac{(V_{FS} - V_{OS}) \times 2^N}{(2^N - 1) \times V_{REF}} - 1\right] \times 100\% \qquad (9.24)$$

for a unipolar converter.

9.13.4 Integral nonlinearity

The integral nonlinearity or **integral linearity error** at a particular code is the deviation of the measured analogue input from the corresponding value calculated from the straight line between the two end points. It may be quoted in volts or in LSB. The specification will include a value for the 'maximum linearity error' which usually occurs at the most significant bit transition.

9.13.5 Differential nonlinearity

Differential nonlinearity or **differential linearity error** or **step error** is the difference between the change of analogue input needed for 1 LSB change in output and the calculated value. Note that in a successive approximation converter, nonmonotonicity in the D/A causes missing code(s) in the A/D; that is, certain codes never appear at the output no matter what the value of the input.

9.13.6 Dynamic tests

Static tests are inadequate for A/Ds which are to be used at high conversion rates such as in audio and video systems. They are also too slow for routine testing by either manufacturers or users of ICs. A number of specialized tests as described below have been developed and the equipment required is commercially available (McLeod, 1991; Sheingold, 1986; Hewlett-Packard, undated).

9.13.7 Back-to-back test

The A/D under test and a D/A are connected back to back. The analogue test input may be a ramp or a sinewave. The difference between it and the output of the D/A is amplified and displayed on a CRO with the test signal providing the time base. Obviously, the D/A and the differential amplifier must have less error than the A/D under test.

9.13.8 Fast Fourier transform test

The test signal is a pure sinewave at whatever frequency is appropriate for the A/D. The converter is accurately triggered by a crystal-controlled clock. The resulting set of samples is stored in the computer as a function of time and the fast Fourier transform (FFT), which is explained in more detail in Chapter 12, is used to analyse it as a function of frequency. Ideally the output of the FFT comprises quantization noise uniformly spread throughout the spectrum and a single large spike at the sampling frequency. Nonlinearity errors show up as spikes at frequencies which are harmonics of the sampling frequency. Quantization noise is discussed further in Chapter 10.

9.13.9 Beat frequency test for fast converters

Testing video converters requires special apparatus and techniques, and parameters measured in low-speed tests do not necessarily apply when a converter is working at high rates (over a million samples per second).

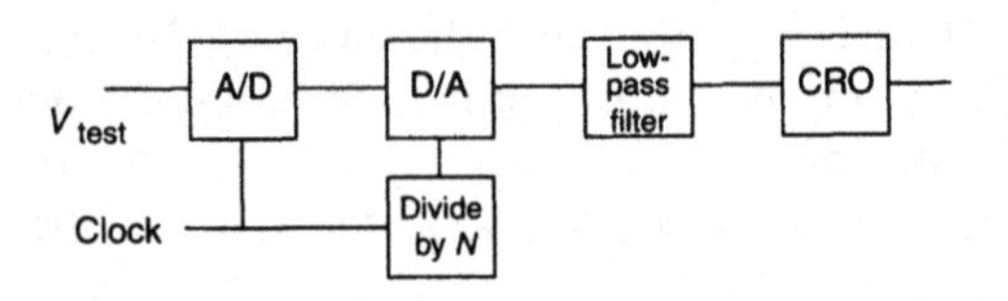

Figure 9.17 *Beat frequency test.*

The beat frequency test is illustrated in Figure 9.17. The input signal is a high-frequency, pure sinewave. The A/D is converting at a rapid rate, but only one in every N samples is reconverted to analogue form, so that a slower and more accurate D/A may be used. The sampling frequency is slightly different from the signal frequency and, ideally, the interval between samples is one signal period + the time to change by 1 LSB. The frequency of the output of the D/A is the beat frequency between them. Sooner or later every level will be digitized and displayed. Quantization noise is removed by the low-pass filter. Missing codes and spurious codes are seen on the CRO.

9.13.10 Histogram test for fast converters

The test input signal is a full-range, high-frequency sinewave. All the output codes are stored in a high-speed RAM and subsequently analysed by a computer. The histogram shows how many times each code has been recorded. Missing codes and differential nonlinearities become obvious.

9.13.11 Other tests

Tests may also be performed for signal-to-noise ratio, and for the effects of changes in the supply voltage and the temperature on the other parameters.

9.14 Examples

Example 9.1

This example demonstrates that the frequency of a signal which can be tracked accurately throughout its cycle is surprisingly low, especially if the signal utilizes the full range of the converter and the resolution of the converter is high.

A 10-bit tracking converter incorporates a 1.235 V reference, and the clock frequency is 100 kHz. Calculate:

1. The maximum slewing rate of input signal that the converter can track.
2. The maximum frequency of a full range, sinusoidal signal that the converter can track.

Solution

1. The maximum slewing rate that a tracking converter can follow is 1 LSB per clock period:

$$\text{LSB} = \frac{1.235}{1024} = 1.206\ \text{mV}$$

$$\text{Maximum slewing rate} = \frac{0.001\,206}{10^{-5}} = 120.6\ \text{V s}^{-1}$$

2. The converter range is $V_{REF} - 1$ LSB. Maximum amplitude A which can be digitized is $\frac{1}{2}V_{REF} - 1$ LSB = 0.6163 V.
   The maximum slewing rate of a sinusoidal signal is $2\pi fA$. Therefore, the maximum frequency sinusoid which the A/D can follow at maximum amplitude is given by

$$2\pi f \times 0.6163 = 120.6$$

$$f = 31.14\ \text{Hz}$$

Example 9.2

Examples 9.2 and 9.3 examine the effect on the accuracy of an integrating converter when the supposedly constant input is subject to mains frequency interference.

A 4½-digit dvm is being used to measure a voltage whose mean value is one volt. Calculate the maximum amplitude of mains frequency hum which can be superimposed on this voltage without causing more than ± 1 LSD variation in the reading when the actual mains frequency is 50.2 Hz and the integration period is 200 ms. What is the SMRR under these conditions?

Solution

One volt measured on a 4½-digit meter reads 1.0000. Therefore, 1 LSD represents 0.1 mV:

$$\pi f T_{INT} = \pi \times 50.2 \times 0.2 = 31.5416 \text{ rad}$$

and

$$\sin(\pi f T_{INT}) = 0.125\,333$$

From (9.4) the series mode rejection ratio is

$$\text{SMRR} = \frac{\pi f T_{INT}}{\sin(\pi f T_{INT})}$$

$$= \frac{31.5416}{0.125\,333}$$

$$= 251.662 \quad \text{or} \quad 48 \text{ dB}$$

The maximum amplitude of hum which can be tolerated on the input if the error is not to exceed 1 LSD, i.e. 0.1 mV, is

$$\text{SMRR} \times 0.1 = 251.662 \times 0.1$$

$$= 25.2 \text{ mV}$$

Note that we use the full precision value of SMRR in the calculation and then round the answer to the nearest 0.1 mV, as that is the resolution of the meter in question. The input is 1 V, so the maximum amplitude of hum must not exceed 2.5% of input.

Example 9.3

A 4½-digit voltmeter has an integration period of 100 ms exactly, but the mains frequency may vary in the range $(50 \pm x)$ Hz, where $x << 50$. Mains frequency noise having an amplitude of 1% of full-scale voltage is added to the input voltage. Calculate the maximum value of the frequency deviation x such that the reading does not fluctuate by more than ± 1 LSD.

Solution to Example 9.3

A 4½-digit voltmeter has four digits which can display any number from 0 to 9 and one which can display 0 or 1 only. Thus its full-scale reading is 19 999, and 1% of reading is 200 LSD.

$$\text{Minimum SMRR} = \frac{\text{noise}}{\text{maximum allowable error}}$$

$$= \frac{\pm 200}{\pm 1} = 200$$

From (9.4) SMRR at the limiting frequencies is

$$\text{SMRR} = \frac{\pi f \tau}{\sin(\pi f \tau)}$$

The frequency limits are $f = 50 \pm x$ Hz and the integration time $\tau = 0.1$ s. Therefore,

$$200 = \frac{0.1\pi(50 \pm x)}{\sin(0.1\pi(50 \pm x))}$$

$$= \frac{5\pi \pm 0.1\pi x}{\sin(5\pi)\cos(0.1\pi x) \pm \sin(0.1\pi x)\cos(5\pi)}$$

Now, $\cos(5\pi) \approx -1$ and $\sin(5\pi) \approx 0$. Also $x << 50$, so $\sin(0.1\pi x) \approx 0.1\pi x$. This gives

$$200 = \frac{5\pi \pm 0.1\pi x}{-(\pm 0.1\pi x)}$$

$$= -\left(1 \pm \frac{50}{x}\right)$$

and $x = \pm\, 0.249$ Hz. Therefore, the frequency limits are $50 \pm 0.249 = 49.751$ and 50.249 Hz.

Example 9.4

This problem deals with the combination of common mode rejection which was discussed in Chapter 4 and series mode rejection which was discussed here in Chapter 9.

Figure 9.18 shows the equivalent input circuit of a dvm.

1. Calculate the true common mode rejection ratio at d.c. and at 49 Hz.

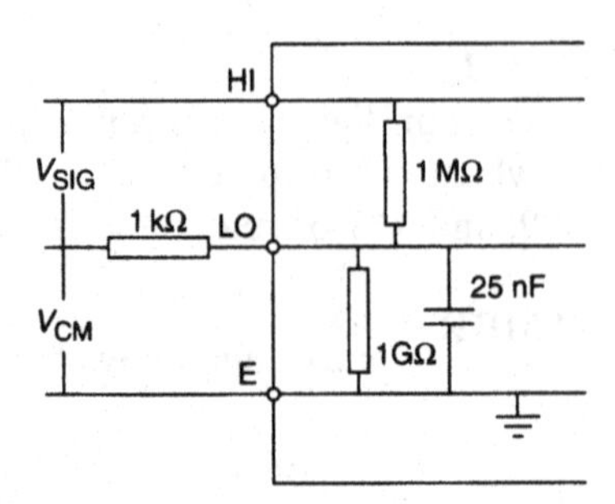

Figure 9.18 *Example 9.4.*

2. What is the maximum common mode voltage (V_{CM}) which can be applied without affecting the reading when the dvm is on a 2 V a.c. range? The resolution of the dvm is 4½ digits.
3. Calculate the series mode rejection ratio of the dvm at d.c. and at 49 Hz if the integration time is 100 ms.
4. What is the effective CMRR of this dvm at d.c. and at 49 Hz when it is on a d.c. range?

Solution

1. Redrawing the equivalent circuit with $V_{SIG} = 0$ and no source resistance gives Figure 9.19. The effective resistance of 1 kΩ + 1 MΩ in parallel is

$$\frac{10^6 \times 10^3}{10^6 + 10^3} = \frac{10^6}{1001} = 999\ \Omega$$

At d.c.,

$$V_{HL} = \frac{V_{CM} \times 999}{10^9 + 999}$$

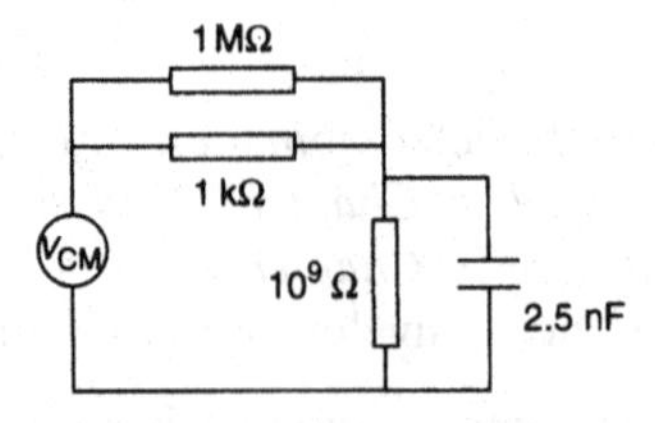

Figure 9.19 *Example 9.4 solution.*

$$\text{CMRR} = \frac{V_{CM}}{V_{HL}} = \frac{10^9 + 999}{999} = 1\,001\,002 \equiv 120\text{ dB}$$

In general, reactance of 0.0025 μF at f Hz is

$$\frac{-j}{2\pi f \times 2.5 \times 10^{-9}} = \frac{-j\,63.662}{f}\text{ M}\Omega$$
$$= 1.2992\text{ M}\Omega \text{ at } 49\text{ Hz}$$

This is much less than 1 GΩ so the effective impedance from LO to ground is $-j1.2992$ MΩ.

$$V_{HL} = \frac{jV_{CM} \times 999}{1.2992 \times 10^6}$$

Taking the modulus of the ratio V_{CM}/V_{HL} gives

$$\text{CMRR} = \frac{1.2992 \times 10^6}{999} = 1300.5 = 62.28\text{ dB}$$

2. A 4½-digit meter reads up to 19 999, so on the 2 V range the maximum reading is 1.9999 V and a least significant digit represents 100 μV.

   The a.c. common mode voltage which would produce 100 μV at the meter terminals is (to reasonable accuracy)

$$100 \times 10^{-6} \times 1300.5 = 0.13\text{ V}$$

3. The series mode rejection ratio (9.4) is

$$20\log_{10}\frac{\frac{1}{2}\omega\tau}{\sin\left(\frac{1}{2}\omega\tau\right)}$$

At d.c., SMRR = 20log 0/0, which is underfined but, as ω tends to 0, the limit of

$$\frac{\frac{1}{2}\omega\tau}{\sin\left(\frac{1}{2}\omega\tau\right)}$$

is 1, and log 1 = 0. Therefore, at d.c., SMRR = 0.

At 49 Hz,

$$\tfrac{1}{2}\omega\tau = 49\pi\tau = 15.39$$

$$\sin(\tfrac{1}{2}\omega\tau) = 0.3090$$

$$\text{SMRR} = 20 \log_{10} \frac{15.39}{0.309} = 49.80 = 33.95 \text{ dB}$$

4. When the meter is on a d.c. range, an a.c. common mode voltage produces a reduced a.c. series mode voltage at the meter terminals, and integration further reduces the reading this causes.

   From equation (4.4), on a decibel scale,

   $$\text{effective CMRR} = \text{true CMRR} + \text{SMRR}$$

   At d.c.,

   TCMRR = 120 dB

   SMRR = 0 dB

   ECMRR = 120 dB

   At 49 Hz,

   TCMRR = 62.28 dB

   SMRR = 33.95 dB

   ECMRR = 106.23 dB

   Note that all these calculations have assumed that the unwanted common and series mode signals are not so large as to drive the input circuits out of their linear working ranges. This can happen and then the readings become completely unreliable.

Example 9.5

To work accurately, a dual ramp converter chip must be associated with the correct passive components. This example shows how to calculate their values, and illustrates the working of the analogue auto-zero correction circuit.

A dual ramp converter has a maximum reading of 2.9999 V. The power supply to the integrator is ± 15 V, and the integration time is 100 ms. The input resistor to the integrator is 200 kΩ. The input bias current is 50 pA and the input offset voltage is 2 mV.

1. If there is no auto-zero correction, what is the reading when the input terminals are short-circuited?
2. Find the smallest standard value of integrator capacitor which can be used without overloading the integrator.
3. An auto-zero circuit is added. When the switch is open, it has a self-capacitance which is effectively 4 pF to the common of

the supply. What is the minimum value of the auto-zero capacitor if the charge suck-out is to be equivalent to less than 1 LSD?

Solution

1. Integrator offset error is

$$2 \times 10^{-3} + 50 \times 10^{-12} \times 2 \times 10^{5}\ \text{V} = 2.01\ \text{mV}$$

The maximum reading is 2.9999 V, therefore, LSD represents 100 μV.

$$2.01\ \text{mV} \equiv 20\ \text{LSD} \equiv 20\ \text{clock pulses}$$

2. The operational amplifier supply voltage is ± 15 V. Therefore, a reasonable swing for the output voltage V_A would be ± 13 V. The minimum size of the integrating capacitor is given by

$$\frac{1}{RC} = \int_0^{T_1} V_{IN}\, dt \leqslant 13$$

Hence,

$$\frac{2.9999 \times 0.1}{200 \times 10^3 \times C_{INT}} \leqslant 13$$

$$C_{INT} \leqslant 0.116\ \mu\text{F}$$

The next larger standard size capacitor would be 0.15 μF.

3. When the auto-zero switch opens, the charge on the auto-zero capacitor is shared with the self-capacitance of the switch, so the charge is reduced by the fraction

$$\frac{C_S}{C_S + C_{AZ}}$$

Now, the auto-zero voltage is 2.01 mV, and should not be reduced by more than 100 μV, i.e. the equivalent of 1 LSB:

$$\frac{C_S}{C_S + C_{AZ}} \leqslant \frac{0.1}{2.01}$$

$$\frac{C_S + C_{AZ}}{C_S} \geqslant 20.1$$

$$C_{AZ} \geqslant 19.1C_S$$

$$C_{AZ} \geqslant 76.4 \text{ pF}$$

During the integrate and deintegrate phases, charge is being removed from the auto-zero capacitor by the integrator bias current. If the capacitor is too small, the auto-zero voltage will not adequately cancel the offset at the crucial moment at the end of the deintegrate period when the comparator operates. The integration time is 100 ms, and the deintegration time for maximum input voltage would be 200 ms, so the auto-zero capacitor may need to hold its voltage for 300 ms. The integrator bias current is 50 pA, and the voltage equivalent to 1 LSB is 100 μV.

The change of charge on the auto-zero capacitor during integration and deintegration is

$$50 \times 10^{-12} \times 0.3 \leqslant C_{AZ} \times 10^{-4}$$

Therefore, $C_{AZ} \geqslant 0.15$ μF.

Thus, using an auto-zero capacitor of 0.15 μF, the bias current would change the auto-zero voltage by the equivalent of 1 LSB, but the effect of charge sharing would be trivial. A capacitor of 0.3 μF would be suitable. In practice, operational amplifiers are available with a wide range of offset voltages and bias currents, which are both very sensitive to temperature. It would be preferable to choose an amplifier with a bias current much smaller than the example given.

Example 9.6

A high-accuracy dvm includes a microprocessor and a nonvolatile memory. In the calibration mode the A/D converter readings for zero and 10.000 00 V are stored in memory.

In the measuring mode, the readings for zero and the unknown voltage are stored. The unknown voltage is then computed and displayed with corrections for offsets and sensitivity.

How many bits are required to represent all possible readings of a 6½-digit dvm in offset binary code?

Table 9.2 shows the codes stored in memory after a reading. Complete the table.

Table 9.2 *Data for Example 9.6*

Code stored	*Hexadecimal*	*Decimal*
Nominal values		
Zero	20 0000	?
Maximum positive	3F FFFF	?
Maximum negative	00 0000	?
Calibration values		
Zero	20 0040	?
10.000 00 V	30 0100	?
Code change for 10.000 00 V	?	?
Measurement values		
Zero	20 0660	?
Unknown voltage	37 5ABC	?
Code change for unknown voltage	?	?
Displayed voltage	?	?

Calculate the measured voltage. You may assume perfect linearity.

Solution

The maximum reading of a 6½-digit meter is ± 1 999 999. The next larger binary number is

$$2^{21} = 2\,097\,152$$

To represent 6½ digits in binary format requires 21 bits + sign bit = 22 bits. The completed table (Table 9.3) was obtained using the following values for powers of 16 is:

$$\begin{aligned} 16^0 &= 1 \\ 16^1 &= 16 \\ 16^2 &= 256 \\ 16^3 &= 4\,096 \\ 16^4 &= 65\,536 \\ 16^5 &= 1\,048\,576 \end{aligned}$$

Table 9.3 *Auto-zero data for Example 9.6 (completed version of Table 9.2)*

Code stored	*Hexadecimal*	*Decimal*	
Nominal values			
Zero	20 0000	$2 \times 1\ 048\ 576$	$= 2\ 097\ 152$
Maximum positive	3F FFFF	$4 \times 1\ 048\ 576 - 1$	$= 4\ 194\ 303$
Maximum negative	00 0000	0	
Calibration values			
Zero	20 0040	$2\ 097\ 152 + 64$	$= 2\ 097\ 216$
10.000 00 V	30 0100	$3 \times 1\ 048\ 576 + 256$	$= 3\ 145\ 984$
Code change for 10.000 00 V		$3\ 145\ 984 - 2\ 097\ 216$	$= 1\ 048\ 768$
Measurement values			
Zero	20 0660	$2\ 097\ 152 + 6 \times 256 + 6 \times 16$	$= 2\ 098\ 784$
Unknown voltage:	C	12×1	$= 12$
	B0	11×16	$= 176$
	A00	10×256	$= 2\ 560$
	5 000	5×4096	$= 20\ 480$
	70 000	$7 \times 65\ 536$	$= 458\ 752$
	300 000	$3 \times 1\ 048\ 576$	$= 3\ 145\ 728$
	375 ABC	total	$= 3\ 627\ 708$

Code change for unknown voltage $3\ 627\ 708 - 2\ 098\ 784 = 1\ 528\ 924$

$$\text{Measured voltage} = \frac{10 \times 1\ 528\ 924}{1\ 048\ 767} = 14.578\ 286$$

Displayed voltage reading (6½-digit resolution) + 14.578 29

Example 9.7

1. A charge-balance converter works on a clock frequency of 20 kHz. What pulse rates are generated at maximum positive input voltage, maximum negative input voltage and zero input voltage?
2. What integration time is required for 10-bit resolution?
3. What uncertainty can be allowed in the pulse length if the error due to this must not exceed ¼ LSB?
4. How could the clock frequency be modified to improve the rejection of 50 Hz noise?

Solution

1. When the input is maximum positive, every clock pulse switches the reference current either off or on, so the pulse rate is $\frac{1}{2}f_c$, i.e. 10 kHz. When the input is maximum negative, the input current is equal to the offset current and so the pulse rate is zero. Zero input voltage generates a pulse rate of 5 kHz.
2. Ten-bit resolution requires 2^{10}, i.e. 1024 pulses in the integration time. 1024 pulses at 10 kHz takes 0.1024 s. ¼ LSB is an error of 1 part in 4096. Nominal pulse duration is

$$1/20\,000 \text{ s} = 50\ \mu\text{s}$$

3. Uncertainty in pulse duration is

$$\frac{50}{4096}\ \mu\text{s} = 12.2 \text{ ns}$$

4. For improved rejection of mains frequency noise, the integration period must be a multiple of the mains period. Thus rejection would be improved if the integration period were changed from 0.1024 s to 0.1000 s. The corresponding clock frequency would be

$$\frac{0.1024}{0.1000} \times 2000 = 20.48 \text{ kHz}$$

Example 9.8

This example is part of the design of a meter to measure noise. To measure 'noise pollution level', the noise is filtered to the same frequency range as the human ear, and measured in decibels, which is a logarithmic scale. This signal is then averaged over one-second intervals. A short, loud noise is rated as equally objectional to a quieter, continuous noise. A charge-balance converter was used as an integrator because the input is connected all the time. If a dual-ramp converter had been used, any loud bangs which occurred in the deintegrate periods would have been missed.

A synchronous charge balance converter is to be used to measure a varying input voltage repeatedly averaged over one-second intervals.

If the input range is 0–10 V, and the counter has 16 bits, calculate the clock frequency for best resolution and the smallest change in input voltage which can be measured.

Solution

The maximum input voltage is determined by the output frequency which fills the counter during the gate time. In this example, the maximum output frequency is 2^{16} counts per second = 65 536 Hz.

For maximum resolution, this must be produced by an input of 10 V. The clock frequency is twice the maximum output frequency = 131.072 kHz. The smallest change in input voltage which can be measured corresponds to

$$1 \text{ LSB} = \frac{10}{2^{16}} \text{ V} = 0.153 \text{ mV}$$

Example 9.9

A 4½-digit voltmeter using charge balance conversion has an integration period of one second. What is the clock frequency and what is the maximum delay in the switching which can be tolerated without causing more than one least significant digit error?

Calculate the amplitude of the series-mode noise at a frequency of 12.5 Hz which would alter the reading by ± ½ LSD.

Solution

The theoretical maximum output frequency is half the clock frequency. This corresponds to $V_{IN} = \frac{1}{2}V_{REF}$. The practical maximum is slightly less: 4½-digit resolution implies 19 999 counts in the integration period of 1 s. Thus,

$$\tfrac{1}{2}f_c = 20\,000 \text{ Hz}$$

Hence, the maximum clock frequency $f_c = 40\,000$ Hz.

Let the switching delay be δt, then the capacitor is discharged for

$$\frac{1}{f_c} + \delta t, \quad \text{instead of } \frac{1}{f_c}$$

The charge removed is increased by the fraction $f_c \delta t$. Now, this error has to be less than 1 in 20 000, i.e. $\leqslant 5 \times 10^{-5}$:

$$f_c \delta t \leqslant 5 \times 10^{-5}$$

$$40\,000 \times \delta t \leqslant 5 \times 10^{-5}$$

$$\delta t \leqslant 1.25 \text{ ns}$$

The error due to series-mode ripple (9.4) is

$$\frac{V_N \sin(\frac{1}{2}\omega\tau)}{\frac{1}{2}\omega\tau} = \frac{V_N \sin(\pi f \tau)}{\pi f \tau}$$

where V_N is the ripple amplitude, ω the ripple frequency and τ the integration period.

On the 2 V range, the maximum reading is 1.9999, so the LSD represents 100 μV.

The series-mode error is

$$\pm 50 \times 10^{-6} = \frac{V_N \sin(\pi \times 12.5 \times 1)}{\pi \times 12.5 \times 1}$$

The ripple amplitude is

$$V_N = \frac{\pm 50 \times 10^{-6} \times \pi \times 12.5}{\sin(\pi \times 12.5)}$$

$$= \pm 1.96 \text{ mV}$$

CHAPTER 10

Sampled data systems

10.1 Introduction

Having studied the separate components of a data acquisition system, we now consider the errors which are inherent in all sampled data systems. These are sampling and quantization errors. They can be reduced, but not eliminated. Examples of the design of complete systems are considered in which these errors are reduced to a level which is tolerable for each particular application.

10.2 Sampling

Sampling breaks the continuously varying analogue input signal or signals into a finite number of **data points** which can be processed by a computer. Information about the signal in between the points is lost for ever, and it is therefore very important that sufficient points are recorded. As more data points necessitate more memory space and faster acquisition, this is yet another example of the trade-off between quality and cost. The signal is not only sampled in time, but its voltage is quantized into a finite number of levels. The system designer must choose the optimum sized quanta for each.

10.3 Quantization

Each data point can have only one of a limited numbers of levels. In a binary system using N bits, there are 2^N levels, each separated by one least significant bit. Thus each level represents a range of voltages from half a least significant bit below to half a least significant bit above the nominal value. The voltage may be anywhere in that range, so that every data point has an uncertainty of ± ½ LSB. This is the **quantization error**, and it can have

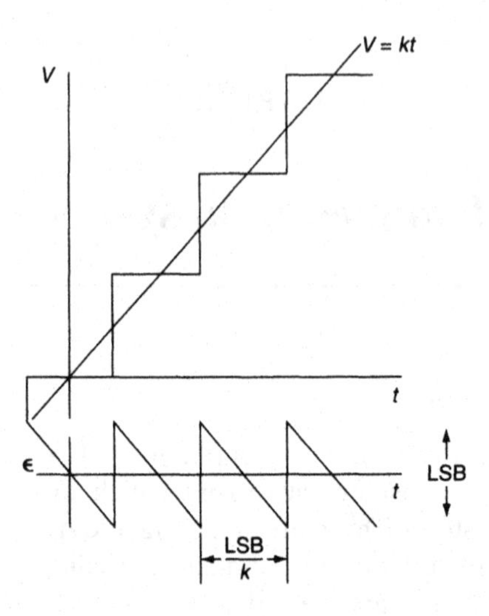

Figure 10.1 *Quantization noise.*

any value from –½ LSB to + ½ LSB. Quantization error is present in all digital measurements even if the A/D converter is perfect.

Figure 10.1 shows how quantization can introduce noise into the output of a D/A converter. Consider a voltage rising at a steady rate $V = kt$, digitized and reconverted to analogue form. The output of the D/A is the staircase waveform shown. The difference between the ramp and the staircase is the quantization error of the A/D and the quantization noise ε in the reconstituted waveform.

10.3.1 RMS quantization error

Consider one period of the sawtooth error waveform from $t = -½k$ to $t = +½k$. The equation of the error waveform is $\varepsilon = -kt$, and the error changes from –½ LSB to +½ LSB:

$$\varepsilon_{\text{rms}} = \left[k \int_{-(1/2)k}^{+(1/2)k} (-kt)^2 \,\mathrm{d}t \right]^{1/2} \tag{10.1}$$

$$= \left[\frac{k}{3} \left(k^2 t^3 \right)_{-(1/2)k}^{+(1/2)k} \right]^{1/2}$$

$$= \left(\frac{1}{24} + \frac{1}{24}\right)^{1/2}$$

$$= \left(\frac{1}{12}\right)^{1/2}$$

$$= 0.289 \text{ LSB}$$

10.3.2 Signal to quantization noise ratio

The peak-to-peak value of a full-range sinusoidal signal is 2^N LSB and its rms value is $2^N/(2\sqrt{2})$. Thus, the signal-to-noise ratio is

$$\frac{(2^N \sqrt{12})}{(2 \sqrt{2})} = 2^N \sqrt{1.5}$$

or, in decibels,

$$20\,(N \log 2 + \tfrac{1}{2} \log 1.5) = 6.02\, N + 1.76 \text{ dB} \qquad (10.2)$$

Obviously, for smaller signals, the ratio is smaller, which is worse.

10.4 Tracking errors

For the best representation of the input waveform, the sampling frequency should be sufficiently high that the input never changes by more than 1 LSB in each sample period.

As calculated in Chapter 5, for a full-range sinusoidal signal,

$$\text{maximum slewing rate} = A \times 2\pi f_{\text{sig}} \qquad (10.3)$$

where A is the amplitude and f_{sig} is the frequency of the signal.

For a full-range signal, $2A = 2^N$ LSB. Thus,

$$f_{\text{SAM}} \geqslant f_{\text{sig}} \pi 2^N \qquad (10.4)$$

It is usually possible to obtain an adequate representation of the input with a lower sampling frequency, but if the resulting samples are applied to a D/A converter, then it will be necessary to follow the D/A with a low-pass filter to smooth out the steps.

10.5 Aliasing errors

10.5.1 Theory

A more serious error can occur due to the phenomenon of **aliasing**, which results in a set of samples which appear to have come from

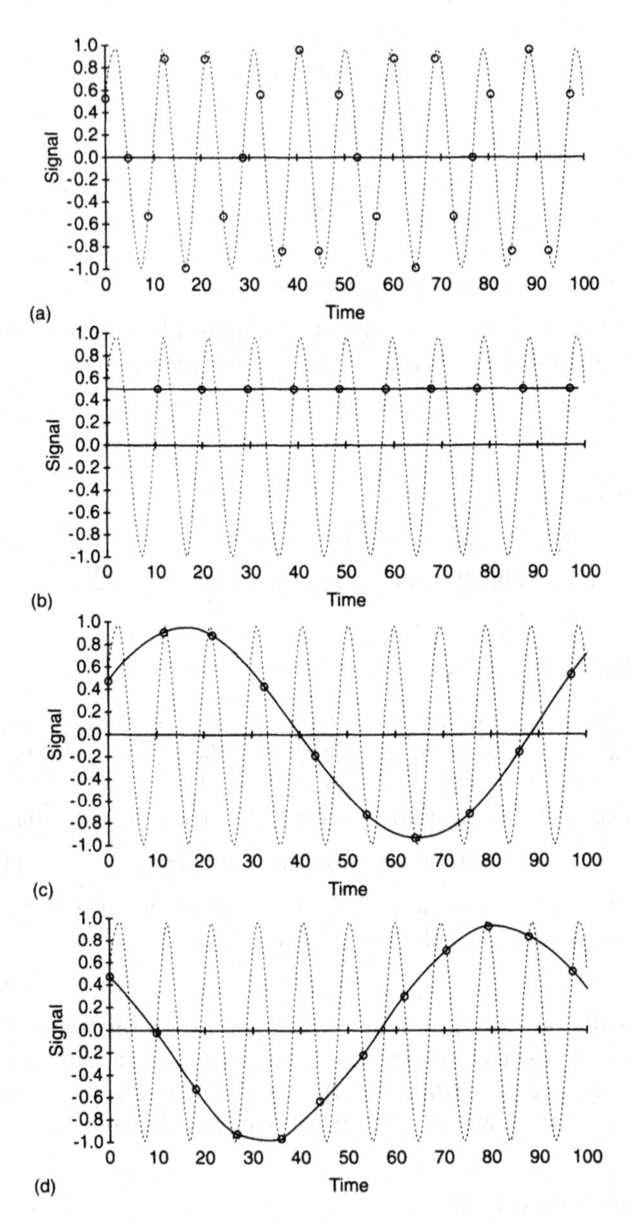

Figure 10.2 *Aliasing in the time domain: (a)* $N > 2$*, correct sampling; (b)* $N = 1$*, sampling frequency eliminated; (c)* $N < 1$*, aliased signal; (d)* $1 < N < 2$*, negative aliased frequency, (······) original signal; (——) aliased signal.*

a signal of a completely different frequency from the real one. This is illustrated in Figure 10.2 in which the sample values are shown by circles. When the only information available is the sample values, it is reasonable to assume that the original signal was that shown by the dashed line. However, the same set of samples might equally well have been obtained from the signal shown by the solid line, or, for that matter, an infinite number of other frequencies. All these frequencies are said to be 'aliases' of each other. Once the signal has been sampled there is no way of knowing what happened between samples, and we therefore assume that the original signal frequency was the lowest which could have produced that set of samples.

To find the values of the aliases, consider a signal

$$y = A \sin(2\pi f_{sig} t + \phi) \tag{10.5}$$

Let the sampling frequency f_{SAM} be Nf_{sig}, i.e. N samples are taken in each period of the signal.

Sample number 0 is taken when $t = 0$, and sample number 1 is taken when $t = 1/f_{SAM} = 1/Nf_{sig}$. The value of sample number 0 is

$$y_0 = A \sin \phi$$

The value of sample number 1 is

$$y_1 = A \sin\left(\frac{2\pi f_{sig}}{Nf_{sig}} + \phi\right)$$

$$= A \sin\left(\frac{2\pi}{N} + \phi\right)$$

Because a sinusoidal function repeats every 2π radians, y_1 may also be written as

$$y_1 = A \sin\left(\frac{2\pi}{N} + 2\pi k + \phi\right) \tag{10.6}$$

where k is any integer, positive or negative. This yields

$$y_1 = A \sin\left(\frac{2\pi(1 + Nk)}{N} + \phi\right)$$

$$= A \sin\left(\frac{2\pi f_A}{Nf_{sig}} + \phi\right)$$

where

$$f_A = f_{sig}(1 + Nk) = f_{sig} + kf_{SAM} \qquad (10.7)$$

The term f_A is called an **alias** of f_{sig}. There are an infinite number of frequencies which are aliases of f_{sig}, one for each value of the integer k. When the only information available is the set of sampled values, we assume that the signal frequency was that represented by joining the points, i.e. the solid line in Figure 10.2. Whether this assumption is correct depends on the number of samples N in each period of the signal. There are four possibilities:

1. More than two samples in each signal period ($N > 2$). Because the sampling is correct (i.e. f_A is minimum when $k = 0$)

$$f_A = f_{sig} \qquad (10.8)$$

   This is the **Nyquist criterion**, and $\frac{1}{2}f_{SAM}$ is called the Nyquist or cut-off frequency.

2. One sample per signal period ($N = 1$). All the samples are equal and the sampled value depends on the phase of the signal at the sampling instants ϕ.

$$f_A = 0 \qquad (10.9)$$

3. Less than one sample per signal period ($N < 1$):

$$f_A = (1 + Nk) = f_{sig} + kf_{SAM} \qquad (10.10)$$

   therefore, k has whatever negative value gives the lowest f_A.

4. More than one, but less than two samples per signal period ($1 < N < 2$). When $k = -1$, $f_A = f_{sig} - f_{SAM}$, which is negative; therefore,

$$f_A = (f_{SAM} - f_{sig}) \qquad (10.11)$$

   Since $\sin(-x) = -\sin(x)$, but $\cos(-x) = \cos(x)$, the negative aliased signal is the complex conjugate of the positive one.

   Figure 10.3 represents the same phenomenon in the frequency domain. The application of the equation $f_A = f_{sig} + kf_{SAM}$ can be demonstrated by drawing the spectrum on a strip of paper and folding it at $\frac{1}{2}f_{SAM}$ and back again at zero and repeating the process until f_{sig} lies between zero and $\frac{1}{2}f_{SAM}$. This is then f_A.

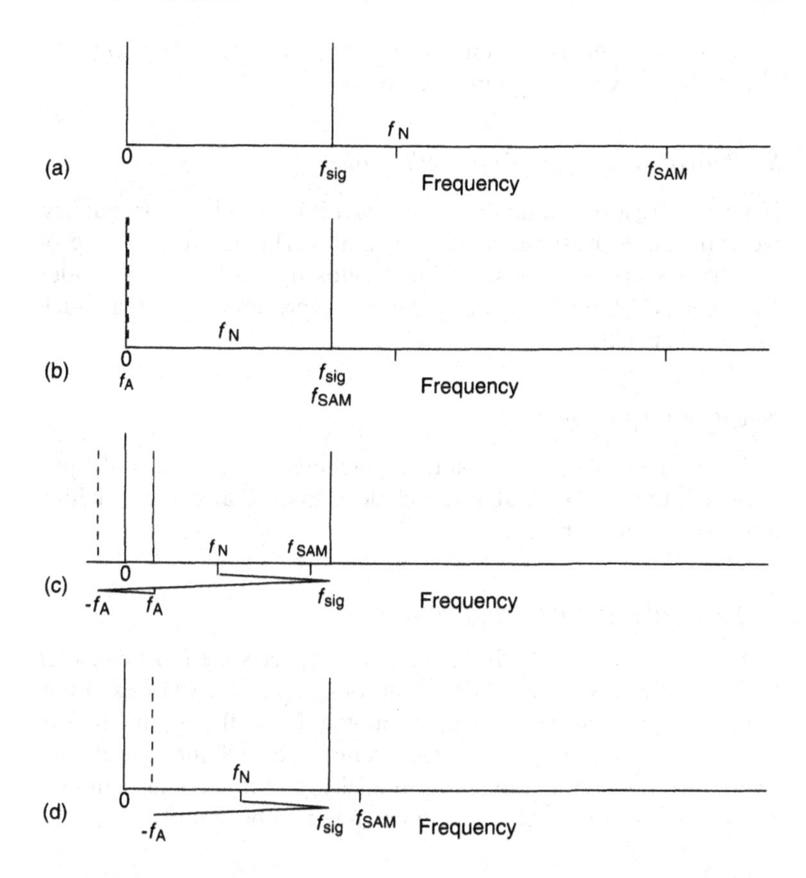

Figure 10.3 *Aliasing in the frequency domain: (a) $N > 2$, correct sampling; (b) $N = 1$, sampling frequency eliminated; (c) $N < 1$, aliased signal; (d) $1 < N < 2$, negative aliased frequency.*

10.5.2 Aliasing errors in practice

Accurate sampling

If we require accurate sampling, we must sample at least twice the frequency of the highest significant component, i.e.

$$f_{SAM} > 2f_{sig} \tag{10.12}$$

Eliminating one frequency component

Components of frequency f_{sig} are removed from the sampled data. For example, to measure the day-to-day temperature variation, we would take measurements at the same time each day and thus eliminate the variation within each day.

Stroboscopes and sampling oscilloscopes

If we require a reproduction of the waveform at a lower frequency, we can sample near the signal frequency. This is the principle of the stroboscope and the sampling oscilloscope. If f_{SAM} is just under f_{sig}, then $N < 1$ and we see a correct representation at a much lower frequency.

Negative frequencies

Using a stroboscope on rotating machinery, if f_{SAM} is only just over f_{sig}, then $N > 1$ but is much less than 2 and the machine appears to rotate backwards.

10.5.3 Choice of sampling frequency

If we have a signal containing many frequencies, we must consider which we need when choosing a sampling frequency. For example, high-frequency noise may be removed from the signal before sampling by a low-pass (anti-aliasing) filter. Some waveforms include components of many frequencies, and it becomes a matter of judgement how many harmonics need to be sampled.

For example:

- square wave, amplitude $\pm A$:

$$y = \frac{4A}{\pi}\left[\frac{\cos \omega t}{1} - \frac{\cos 3\omega t}{3} + \frac{\cos 5\omega t}{5} + \ldots\right] \tag{10.13}$$

- triangular wave, amplitude $\pm A$:

$$y = \frac{8A}{\pi^2}\left[\frac{\cos \omega t}{1} + \frac{\cos 3\omega t}{3^2} + \frac{\cos 5\omega t}{5^2} + \ldots\right] \tag{10.14}$$

- sawtooth wave, amplitude $\pm A$

$$y = \frac{2A}{\pi}\left[\frac{\sin \omega t}{1} - \frac{\sin 2\omega t}{2} + \frac{\sin 3\omega t}{3} + \ldots\right] \tag{10.15}$$

- full-wave rectified sinusoid, amplitude $+A$

$$y = \frac{2A}{\pi}\left[1 + 2\frac{\cos 2\omega t}{1\times 3} - 2\frac{\cos 4\omega t}{3\times 5} + \ldots\right] \tag{10.16}$$

10.5.4 Anti-aliasing filters

When the sampling frequency has been decided, the anti-aliasing filters may be designed. Ideally a very sharp cut-off is required at the Nyquist frequency, but filters with sharp cut-off introduce a large phase shift into the pass-band signals. The filter required depends on the frequency separation between the wanted and unwanted components, and whether the waveform of the signal to be sampled must be preserved.

Filter responses were described in Chapter 4, and are summarized as follows:

- Butterworth, which has the flattest response in the pass-band (Figure 4.7).
- Bessel, which has a phase shift proportional to frequency and therefore causes least waveform distortion because all components are equally delayed (Figure 4.8).
- Chebyshev, which has the steepest cut-off, but has ripple in the pass-band (Figure 4.9).

Note that the filters in all channels must be matched, and when all channels are being monitored the anti-aliasing filters must come before the multiplexer. The only exception to this rule is a system in which the multiplexer is left on one channel for many periods of the signal. The programme selector for radio or television is an example of this, but an acquisition system is trying to record what is happening simultaneously on all the input channels.

A digital filter may be used for anti-aliasing, providing it does not generate spikes. Digital filters have the great advantage that the filter clock frequency may be a multiple, say 10 times, of the sampling frequency, and if the sampling frequency is changed for any reason, the filter frequency tracks it (Hoskins, 1985).

Not all acquisition systems require filters, because sometimes the nature of the sensors ensures that the signals have no high-frequency components. If data storage is not a problem, the sampling rate may be increased until the Nyquist frequency is well above all the signal components. This has the effect of removing the quantization noise from the signal bandwidth.

10.6 System specification

In order to design a complete data logging system, we need a specification which should include the information listed below. It is quite common for the specification given to the instrument engineer to be incomplete, and this is why he should understand as much as possible about the process being measured as well as the measurement techniques which are available to monitor it.

The specification should include the following information:

1. number of analogue input channels;
2. whether channels are to be sampled simultaneously or sequentially;
3. the voltage level of each analogue input;
4. the frequency range of signals to be digitized – if they are a.c., whether they are to be rectified and smoothed before digitization;
5. the accuracy required;
6. the destination of the digital output.

10.7 Example of system design

Most of the decisions the designer has to make are illustrated in the following example:

A system is required to monitor pressure and surface temperature variations at eight points in a machine to 1% accuracy. The pressure sensors have a maximum output of 1 mV and the temperature sensors are radiation detectors having a maximum output of 100 mV and a nonlinear scale. Signal frequencies of 400 Hz are expected and higher frequencies (if present) need not be recorded. The data is to be analysed by computer.

10.7.1 Sensors

The first decisions to be made are the choice of sensors. These depend on the balance of sensitivity, frequency response and cost. This is outside the scope of this chapter, although an overview was given in Chapter 3. In the example above, strain-gauge pressure sensors and infrared radiation detectors had been selected, together with the circuits necessary to energize them.

10.7.2 Signal conditioning

Both types of sensor give a d.c. output, so no demodulation will be necessary. The A/D converter will require an input of the order of volts, so amplifiers will be needed to bring the sensor outputs up to this level. If we are certain that all the pressure sensors have identical sensitivity, we can multiplex the sensor outputs and use a common amplifier. However, if we use separate amplifiers in each channel, they can be individually adjusted for gain, the multiplexer can operate on higher signal levels, and offsets in the multiplexer will not be amplified.

10.7.3 Linearity

Strain gauges are linear, that is their output voltage is proportional to strain. Radiation sensors are very nonlinear because the total radiation varies with the fourth power of the absolute temperature. As the temperature rises, more energy is emitted at shorter wavelengths, but below 200 °C it is almost entirely infrared which most sensors cannot detect. This is illustrated in Figure 3.9.

It is possible to partially correct for nonlinearity with analogue circuits such as diode networks or log amplifiers, but the correction may be made more easily in the computer. However, care must be taken that the resolution of the A/D is sufficient. In the example, we are asked for 1% accuracy, and in the absence of any further information we shall take this as 1% of sensor output voltage.

10.7.4 Order of taking samples

The problem does not state whether the samples are to be taken simultaneously or sequentially, but it is reasonable to assume that the two measurements at each point are required at the same time so, in the absence of more information, we will take both measurements from channel 1 simultaneously, then both measurements from channel 2 and so on. In real life it is quite common for the specification to be incomplete, and the instrument engineer must use his experience to fill in the details. The order of taking samples determines the configuration of the system, which is shown in Figure 10.4.

Multiplexers A and B are addressed together, and the control lines of sample and holds A and B are connected together so that

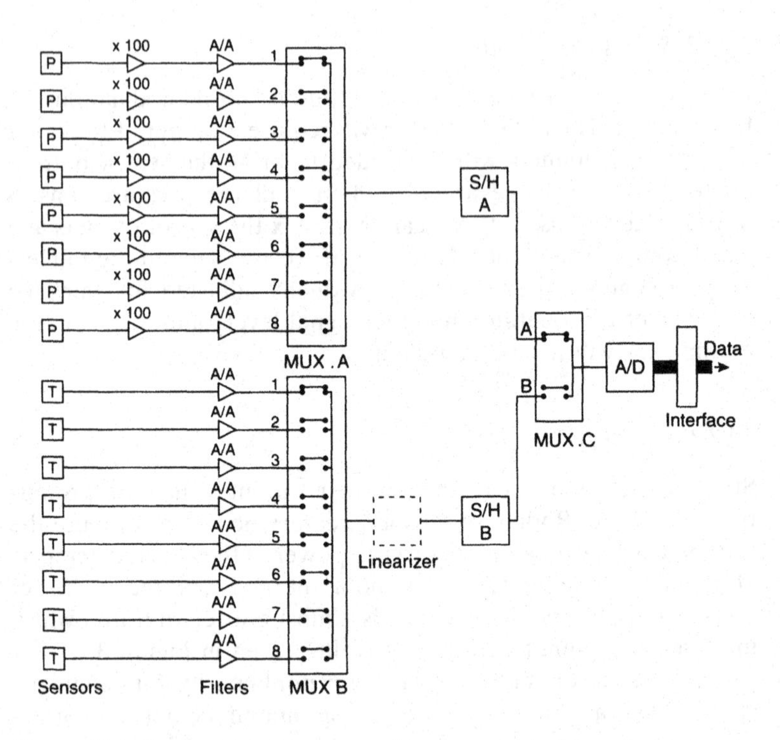

Figure 10.4 *Proposed system to monitor pressure and temperature variations (S/H = sample and hold).*

channel 1 temperature and channel 1 pressure are measured together and so on.

10.7.5 Sampling rate

Signal frequencies up to 400 Hz are expected, so the lowest possible sampling frequency is 800 Hz to avoid aliasing. However, two samples per cycle are not enough to give a good representation of a signal, so a higher frequency is preferable, say 4 kHz. Also, higher frequencies may be present in the signal so anti-aliasing filters should be provided. These must pass the 400 Hz and remove everything above $\frac{1}{2}f_{SAM}$. The signals are sampled at the first multiplexer so anti-aliasing filters have to be placed before the multiplexers.

This sampling rate necessitates solid-state multiplexers and a successive approximation converter.

Table 10.1 *System timing for acquisition units only*

Step	*A/D*	*MUXs A and B*	*S/H*	*MUX C*	*Time required*
0 (initial settings)		channel 1	sample	temp	
1			hold		aperture
2	convert temp 1				convert
3		channel 1 to 2		temp to press	settle
4	convert press 1				convert
5			sample	press to temp	acquire, settle C
6			hold		aperture
7	convert temp 2				convert
8		channel 2 to 3		temp to press	settle C
9	convert press 2				convert
10			sample	press to temp	acquire, settle C
11–35	Repeat for channels 3–8				
36			hold		aperture
37	convert temp 8				convert
38		channel 8 to 1		temp to press	settle C
39	convert press 8				convert
40			sample	press to temp	acquire, settle C
Repeat from step 1					

Total $= 16 \times t_{CON} + 8 \times t_{AP} + 8 \times t_{AQ} + 8 \times t_{SET}$

10.7.6 Timing

The sequence of operation steps may be as in Table 10.1. To complete the table, start with the conversions, then add hold and sample commands, and finally the multiplexer changes. Concurrent operations are written on the same line and the time for the slower operation is underlined in the right-hand column. The table assumes that the time for the multiplexer to change and settle is less than either the conversion time or the acquisition time.

It would be possible to complete the I/O operation during the following conversion provided that the result of each conversion remained latched at the output of the A/D until the following conversion was complete. However, in many A/Ds the output is not available during a conversion, so the read operation must be completed before the start of the next conversion.

When all operations have been fitted into the sequence, the times in the right-hand column are added. This total gives the shortest possible time to acquire one complete set of data correctly.

The price of A/D converters is now so low that it is worth considering using two A/Ds in place of one A/D and a two-channel multiplexer.

The number of samples in a second from one channel is the **sampling rate**, and the total number of samples in a second is the **throughput**:

$$\text{throughput} = \text{number of channels} \times \text{sampling rate}$$

10.7.7 Transferring data to the computer

In addition to controlling the functions of the data acquisition, the computer must read and store each data word which the converter produces. There will normally be time to do this before the next conversion is complete, but it might be necessary to allow additional time for the reading operation(s). Table 10.2 is similar to Table 10.1, with the addition of the computer operations and using the same initial settings. A read operation has been inserted after each conversion followed by write operations controlling the commands which include the S/H control, multiplexer addresses and start conversion. The changes in the addresses of multiplexers A and B can be made at any time after the end of the aperture time, but the control commands are simplified if all multiplexers change together. This is implemented in Table 10.2, and illustrated in Figure 10.5.

Table 10.2 *System timing including communication with computer*

Step	*A/D*	*MUXs A and B*	*S/H*	*MUX C*	*Computer*	*Time required*
0 (initial settings)		channel 1	sample	temp		
1			hold		write	<u>aperture</u>
2	convert temp 1					convert
					read T1	read
3		channel 1 to 2		temp to press	write all MUX	<u>settle</u>
4	convert press 1					convert
					read P1	read
5			sample	press to temp	write MUX C	<u>acquire</u>, settle
6			hold		write	<u>aperture</u>
7	convert temp 2					convert
					read T2	read
8		channel 2 to 3		temp to press	write all MUX	<u>settle</u>
9	convert press 2					convert
					read P2	read
10			sample	press to temp	write MUX C	<u>acquire</u>, settle
11–35	Repeat for channels 3–8					
36			hold		write	<u>aperture</u>
37	convert temp 8					convert
					read T8	read
38		channel 8 to 1		temp to press	write all MUX	<u>settle</u>
39	convert press 8					convert
40			sample	press to temp	write MUX C	<u>acquire</u>, settle
Repeat from first hold						

$$\text{Total} = 16 \times t_{\text{CON}} + 16 \times t_{\text{READ}} + 8 \times t_{\text{AP}} + 8 \times t_{\text{AQ}} + 8 \times t_{\text{SET}}$$

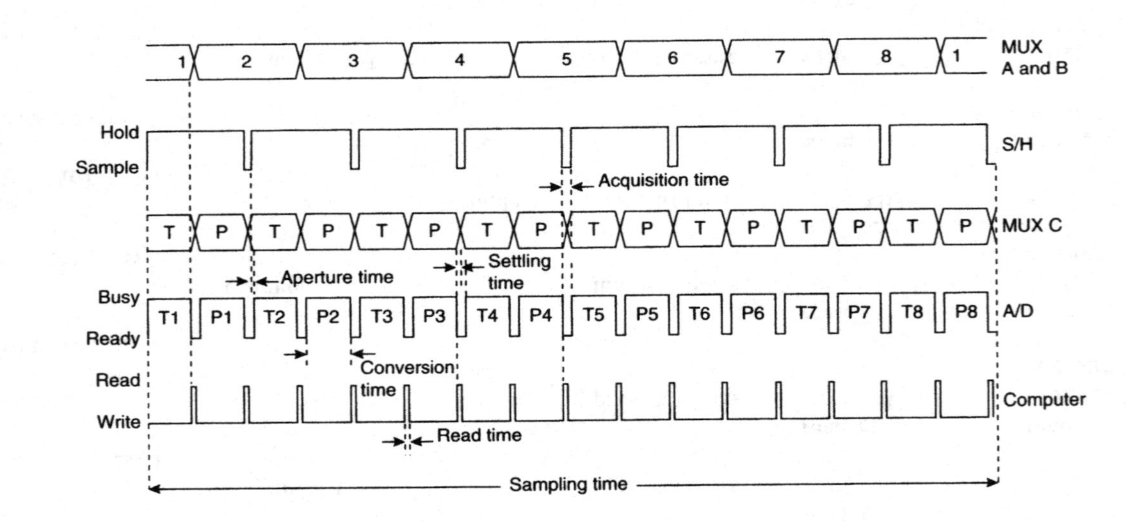

Figure 10.5 *System timing as for Table 10.2.*

Alternatively, a counter timer on the data acquisition board can be used to set the S/H to hold and trigger the A/D. The very short delay (< 1 μs) for the aperture time can be implemented in hardware. However, if the 'start conversion' and 'hold' commands are applied simultaneously, the signal will almost certainly be held and settled before a successive approximation converter decides on the MSB, which is not until one clock period after conversion has started. The 'ready' or 'conversion complete' signal from the A/D can be used to interrupt the computer for a read operation. End-of-conversion can also release the S/H back to sample mode and advance the multiplexer addresses. Of course, if the computer has other tasks to perform in between readings, as it will when operating a real-time control system, then the acquisition process will be slower still.

Further design examples are given at the end of the chapter.

10.8 Control and data signals

At this point it is useful to summarize the signals within a data acquisition system. Table 10.3 lists the units which comprise a system, with their input and output signals. Analogue signals are shown in upper case, and digital signals in lower case.

Table 10.3 *Control and data signals*

Component	*Inputs*	*Outputs*
D/A converter	data REFERENCE	ANALOGUE
A/D converter	ANALOGUE INPUT REFERENCE start enable or chip select byte enable	latched data conversion complete busy/ready or end of conversion
Multiplexer	ANALOGUE address (binary or 1 of N)	SAMPLED ANALOGUE
Sample and hold	ANALOGUE sample to hold control	SAMPLED ANALOGUE

10.9 Error analysis

The many sources of error which contribute to error in each data point are independent of each other, so it is extremely unlikely that they will all be maximum and in the same direction at the same point. To estimate the overall error of the system we take the rms of the errors from all sources.

Each unit in the system contributes its share of errors. As cost rises rapidly with accuracy, the most cost-effective design is one in which all units are of comparable accuracy. To put it another way, there is little point in using a 12-bit converter with a resolution (and hopefully accuracy) of 1 in 4096 if the nonlinearity of the amplifiers is 1%. An error budget is a quantified list of the possible sources of error.

In the example above we have the sensor, amplifier, anti-aliasing filter, multiplexer, sample/hold and converter, making six units in all. The specification calls for 1% accuracy or, to be more precise, 1% uncertainty overall.

$$\varepsilon_{tot} = \surd(\varepsilon_{sen}^2 + \varepsilon_{amp}^2 + \varepsilon_{aa}^2 + \varepsilon_{mux}^2 + \varepsilon_{sh}^2 + \varepsilon_{ad}^2) = 0.01 \qquad (10.17)$$

Thus,

$$\begin{aligned} \text{allowable error per unit} &= \varepsilon_{tot}/\surd 6 \\ &= 0.01/\surd 6 \\ &= 0.00408 \\ &\equiv 1 \text{ in } 245 \\ &\equiv 8 \text{ bits} \end{aligned}$$

In general, to estimate the accuracy required for each unit in a system, divide the overall error allowed in the specification by the square root of the number of units in the system.

The A/D will contribute a maximum quantization error of $\pm \frac{1}{2}$ LSB even if it is otherwise perfect and it is usual to allow ± 1 LSB. The rms quantization error is 0.289 LSB. If an 8-bit converter will suffice, the data can be transferred into the computer in one byte, which saves time, and simplifies the circuit. Eight bits correspond to a resolution of 1 in 256. This should be adequate for 1% overall accuracy with a margin for error in the rest of the system.

10.10 Earthing

Earthing in analogue circuits was discussed in Chapter 4, but the board including the A/D converter has the additional problem that the zero volts line of the digital supply voltage will be at approximately earth potential, but will be noisy because of the short rise and fall times of digital pulses. It is good practice to decouple each integrated circuit with a 0.1 μF capacitor across the power supply as close to the chip as possible.

Always remember that earth leads, tracks and planes have impedance, and that earth is not one vast sink always at zero potential, but that the potential of each nominal earth point depends on the current flowing into it. If digital currents flow in analogue circuits, digital noise becomes series-mode noise in the analogue input.

The low-potential side of the analogue input should not be connected to digital earth. If the analogue low side is nominally at earth potential, analogue and digital earths must be joined together at one point, and one point only, as close to the power supply as possible. If the analogue low side is within a few volts of earth, a differential or instrumentation amplifier in the first stage is the best solution, but if there is a possibility of a common mode voltage large enough to take the input voltage outside the range of the analogue power supply, as in a digital voltmeter, then the analogue and digital circuits must be isolated by a transformer or optocoupler as shown in Figure 4.3.

A comprehensive explanation of these problems is given by Brokaw (1984).

10.11 Examples

Example 10.1

1. What is the quantization noise for a 12-bit D/A converter with a 5.000 V reference?
2. What is the corresponding signal-to-noise ratio for a full range sinusoidal signal?

Solution

1. When $N = 12$, and $V_{REF} = 5.000$ V,

$$\text{noise} = \frac{5.000}{4096 \times \sqrt{12}} = 0.35 \text{ mV rms}$$

2. Signal-to-noise ratio (10.2) is $6.02N + 1.76$ dB. Therefore, when $N = 12$, signal-to-noise ratio = 74 dB.

Example 10.2

A triangular waveform signal of frequency 1 kHz and amplitude ± 1 V is digitized and the resulting samples are applied to an 8-bit digital to analogue converter.

1. Evaluate the minimum sampling frequency to avoid aliasing the fundamental frequency of the signal.
2. Calculate the slewing rate of the signal.
3. What sampling frequency is necessary for the D/A to track the signal with no slew rate error?
4. Sketch the output of the D/A when the sampling frequency is (a) 8 kHz and (b) 1.11 kHz.

Solution

1. To avoid aliasing the fundamental of the triangular waveform, $f_{SAM} \geqslant 2$ kHz.
2. The slewing rate of the signal is 2 V in 0.5 ms = 4000 V s^{-1}.
3. To track accurately, the sampling interval less than or equal to the time to change by 1 LSB:

$$\text{LSB} = 2/2^8 \text{ V}$$

$$t_{SAM} \leqslant \frac{2}{2^8 \times 4000} \text{ s}$$

$$f_{SAM} \geqslant 2^7 \times 4000 = 512 \text{ kHz}$$

4. With a sampling frequency of 8 kHz, the sampled waveform has the same fundamental frequency as the original, but it is not smooth. Increasing the sampling frequency improves the reproduction of the original waveform as more and more of the harmonics in the original triangular waveform are correctly sampled. From equation (10.7), with a sampling frequency of 1.1 kHz, the sampled waveform has aliased fundamental frequencies of $1 \pm k \times 1.1$.

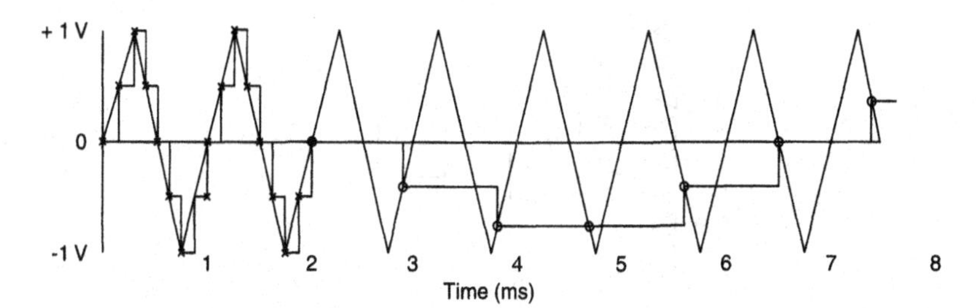

Figure 10.6 *Example 10.2: (×) shows samples at 8 kHz; (○) shows samples at 1.1 kHz.*

The lowest of these is –0.1 kHz, and this is the fundamental of the reproduced waveform.

Example 10.3

This example explores the effect of the layout of the circuit on the maximum sampling rate.

An eight-channel data acquisition system is required and the following components are available:

- 10-bit successive approximation converter, conversion time 20 μs;
- eight-channel multiplexer, settling time (to 0.1%) 10 μs;
- sample and hold units, aperture (including settling) time 0.1 μs, acquisition time 5 μs.

For each of the configurations shown in Figure 10.7, calculate:

1. the minimum time to sample all channels;
2. the maximum signal frequency without aliasing error;
3. the maximum signal frequency without aperture time error.

Solution

The sequences of operations are shown below. If several operations can take place simultaneously, they are shown on the same line, and the longest time is underlined. The time for the slowest operation is shown in the last column.

In Figure 10.7(a) and Table 10.4 the sequence is convert, multiplex, convert, multiplex and so on for eight channels and there are no concurrent operations:

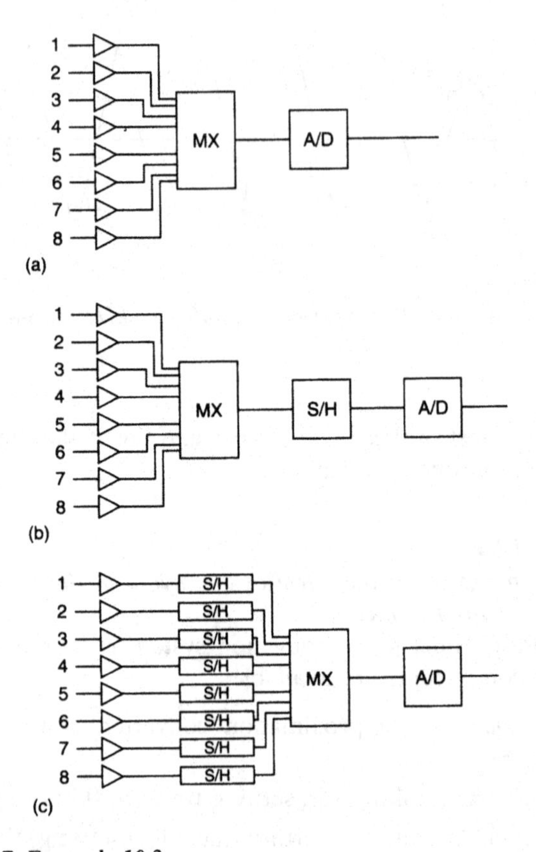

Figure 10.7 *Example 10.3.*

1. The sampling time is 240 μs. The corresponding sampling frequency is 4.17 kHz.
2. The maximum signal frequency to avoid aliasing would be 2.08 kHz.
3. The effective aperture time is the conversion time = 20 μs, so the maximum signal frequency to avoid aperture uncertainty error is

$$\frac{10^6}{1024 \times \pi \times 20} = 15.5 \text{ Hz}$$

which is much less than the aliasing limit.

Table 10.4 *Data acquisition sequences and timing for the circuit of Figure 10.7(a)*

MUX	*A/D*	*Time*	*Actual time (μs)*
	convert 1	t_{CON}	20
To channel 2		t_{SET}	10
	convert 2	t_{CON}	20
To channel 3		t_{SET}	10
	convert 3	t_{CON}	20
⋮	⋮	⋮	⋮
To channel 8		t_{SET}	10
	convert 8	t_{CON}	20
To channel 1		t_{SET}	10
		Total	$8 \times (10 + 20) = 240$

In Figure 10.7(b) and Table 10.5 with the signal from one channel held by the sample and hold, the multiplexer can move on to the next channel and settle while the A/D is converting. The sequence is hold, convert and multiplex, sample, hold, convert and multiplex, sample and so on for eight channels. Conversion takes longer than multiplexer settling so the time allowed must be at least long enough for conversion:

1. The sampling time is 200.8 μs. The corresponding sampling frequency is 4.98 kHz.
2. The maximum signal frequency to avoid aliasing would be 2.49 kHz.
3. The effective aperture time is the aperture time of the sample and hold = 0.1 μs, so the maximum signal frequency to avoid aperture uncertainty error is

$$\frac{10^6}{1024 \times \pi \times 0.1} = 3.11 \text{ kHz}$$

which is more than the aliasing limit.

In Figure 10.7(c) and Table 10.6 all the sample and holds operate simultaneously, and the converter must wait for the multiplexer to settle for each channel. Nevertheless, the multiplexer can return

Table 10.5 *Data acquisition sequences and timing for the circuit of Figure 10.7(b)*

S/H	*MUX*	*A/D*	*Time*	*Actual time (μs)*
Hold	on channel 1		t_{AP}	0.10
	to channel 2	convert 1	$t_{SET}/\underline{t_{CON}}$	20
Sample			t_{AQ}	5
	to channel 3	convert 2	$t_{SET}/\underline{t_{CON}}$	20
Hold			t_{ap}	0.10
	to channel 4	convert 3	$t_{SET}/\underline{t_{CON}}$	20
⋮		⋮	⋮	⋮
Sample			t_{AQ}	5
	to channel 8	convert 7	$t_{SET}/\underline{t_{CON}}$	20
Hold			t_{AP}	0.10
	to channel 1	convert 8	$t_{SET}/\underline{t_{CON}}$	20
Sample			t_{AQ}	5
		Total	$8 \times (0.10 + 20 + 5) = 200.80$	

Table 10.6 *Data acquisition sequences and timing for the circuit of Figure 10.7(c)*

S/H	*MUX*	*A/D*	*Time*	*Actual time (μs)*
Hold	on channel 1		t_{AP}	0.10
		convert 1	t_{CON}	20
	to channel 2		t_{SET}	10
		convert 2	t_{CON}	20
	to channel 3		t_{SET}	10
		convert 3	t_{CON}	20
⋮		⋮	⋮	⋮
	to channel 8		t_{SET}	10
		convert 8	t_{CON}	20
Sample	to channel 1		$t_{AQ}/\underline{t_{SET}}$	10
		Total	$0.10 + 8 \times (20 + 10) = 240.1$	

to its starting point while the sample and holds acquire the signal again ready for the next set of samples. The sequence is hold (convert and multiplex eight times), sample:

1. The sampling time is 240.1 μs. The corresponding sampling frequency is 4.16 kHz.
2. The maximum signal frequency to avoid aliasing would be 2.08 kHz.
3. If the acquisition and multiplexer settling times are not overlapped, 5 μs are added to the time giving a sampling frequency of 4.08 kHz. The effective aperture time and hence the aperture uncertainty error limit are the same as for Figure 10.7(b).

Example 10.4

A procedure for designing a system was given in section 10.7 above. Examples 10.4–10.6 provide further practice.

An oscilloscope may be converted to storage mode by adding a digital data capture system. Design a system for sampling four channels simultaneously. Sketch a block diagram of the layout and explain the sequence of operations. Illustrate your answer with a timing diagram.

Calculate the maximum conversion rate and the maximum signal frequency for which this system could be used, given the following unit specifications:

- analogue to digital conversion time t_{CON} 20 μs
- sample and hold aperture time t_{AP} 150 ± 10 ns
- sample and hold acquisition time t_{AQ} 6 μs
- multiplexer settling time t_{SET} 3 μs
- multiplexer address decoding time t_{ADD} 500 ns
- RAM writing time t_{WT} 300 ns

Solution

The layout of a four-channel data capture system is shown in Figure 10.8. The sequence of operations is shown in Table 10.7. If several operations can take place simultaneously, they are shown on the same line, and the longest time is underlined. The time for the slowest operation is shown in the last column. The control may be simplified by making operations sequential, instead of simultaneous, but the system will then run more slowly.

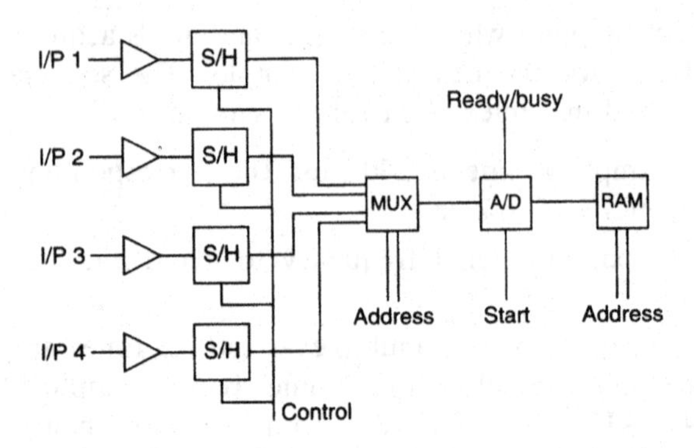

Figure 10.8 *Example 10.4.*

Table 10.7 *Data acquisition sequence and timing for Example 10.4*

S/H	*MUX*	*A/D*	*RAM*	*Time*	*Time*
Hold	on ch. 1			t_{AP}	0.16
		conv. 1		t_{CON}	20
	to ch. 2		write	$(t_{ADD} + t_{SET})/t_{WT}$	3.5
		conv. 2		t_{CON}	20
	to ch. 3		write	$(t_{ADD} + t_{SET})/t_{WT}$	3.5
		conv. 3		t_{CON}	20
	to ch. 4		write	$(t_{ADD} + t_{SET})/t_{WT}$	3.5
		conv. 4		t_{CON}	20
Sample	to ch. 1		write	$t_{ACQ}/(t_{ADD} + t_{SET})/t_{WT}$	6
		Total		$t_{AP} + 4t_{CON} + s(t_{ADD} + t_{SET}) + t_{ACQ} =$	96.66

The sampling frequency corresponding to 96.66 μs is 10.35 kHz, and so the maximum signal frequency to avoid aliasing would be 5.17 kHz.

In practice, allowing for more than two samples per signal period, the upper limit would be closer to 1 kHz.

Example 10.5

1. An 8-bit analogue to digital converter has a range from –10 V to (+10 V – 1 LSB). What is the maximum quantization error?

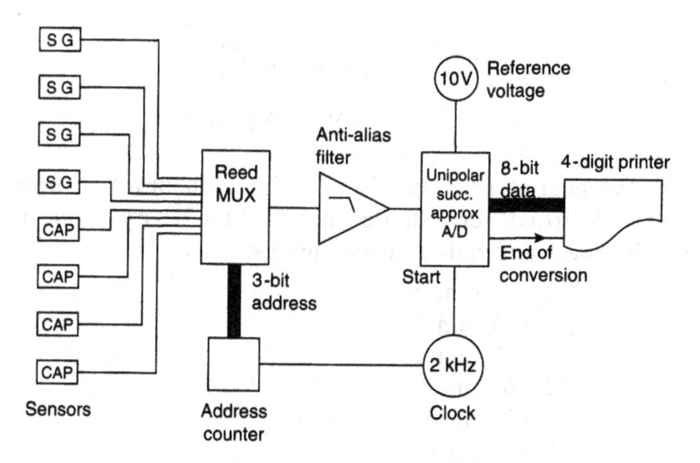

Figure 10.9 *Example 10.5.*

2. What are the digital equivalents of –2.061 V in offset binary and two's complement codes, and what is the quantization error for this input?
3. Figure 10.9 shows a proposed layout for simultaneously monitoring the signals from four strain-gauge transducers with an output of ± 5 mV and from four variable capacitance transducers with an output of 0.7 V rms. There are approximately eight deliberate omissions or errors in the design. What are they and how could the design be improved?

Solution

1. Total range is 20 V – 1 LSB, i.e. = $2^8 - 1$ LSB. Therefore,

$$1\ \text{LSB} = \frac{20}{2^8} = 0.078\,125\ \text{V}$$

Maximum quantization error is ± ½ LSB = ± 39.1 mV.

2. In offset binary code

    0000 0000 represents –10 V

    1000 0000 represents 0 V

$$-2.061 = -10 + 7.939$$

$$7.939\ \text{V} \equiv \frac{7.939}{0.078\ 125}\ \text{LSB}$$

$$= 101.6192\ \text{LSB}$$

To the nearest LSB, this is 102 with a quantization error of –0.3808. To convert 102 decimal to 8-bit binary divide by 2, eight times, and note the remainders:

Half 102 = 51 rem 0
51 = 25 rem 1
25 = 12 rem 1
12 = 6 rem 0
6 = 3 rem 0
3 = 1 rem 1
1 = 0 rem 0
0 = 0 rem 0

The offset binary code for –2.061 V is 0110 0110. To find the two's complement code, invert the MSB, which gives 1110 0110. The quantization error is

$$-0.3808\ \text{LSB} \equiv -0.3808 \times 0.078\ 125 = 0.02975\ \text{V}$$

3. Figure 10.9 has three omissions:

(a) *Signal conditioning*. The strain gauges give a maximum output of ± 5 mV d.c., but the A/D converter has a range of 0–10 V d.c. Differential amplifiers are needed and instrumentation amplifiers would be best. It will be necessary either to offset the output so that the midrange point of the input, which is zero volts, provides the midrange point of the output, which is five volts, or to use a bipolar A/D converter (errors point (c) below).

(b) *Signal conditioning*. The capacitance sensors provide 1 V a.c. and so need demodulators as well as amplifiers. Phase-sensitive demodulators would provide a bipolar output and reject noise and mains hum.

(c) *Sampling*. To ensure simultaneous sampling, sample and hold circuits are needed before the multiplexer.

Figure 10.9 also has five errors:

(a) The clock is too fast for a reed multiplexer and for the printer. The multiplexer switching speed should not be more than 100 channels per second and the printer will print approximately 10 characters per second.

(b) The anti-aliasing filter should be before the multiplexer.

(c) It would be better to use a bipolar A/D converter. This would give a more stable zero than an offset in the amplifier described in omissions point (a) above.

(d) A settling time must be allowed after switching the multiplexer, before starting the conversion. There is nothing in the diagram to show the relative times of the multiplexer address change and the start conversion signals.

(e) An 8-bit converter is not compatible with a 4-digit printer, either in resolution or in coding.

Other considerations are that the sampling rate is not specified. Without knowing this or the application of the system, we cannot say whether the clock is too fast or the multiplexer and printer are too slow. The choice of a printer implies a low sampling speed, i.e. less than 10 samples per second. At this speed reed switches would be suitable, but an integrating converter would be better than successive approximation. If the signals change very slowly, i.e. a slewing rate < 1 digit per millisecond, it might be possible to scan all the channels quickly, and then pause before taking the next set of samples. This would avoid the need for sample and hold circuits; a successive approximation converter would still be needed. On the other hand, if we assume that the clock rate is correct, and the sampling rate is 2000 samples per second, then the multiplexer must be solid state, and the converter successive approximation. At this speed, the printer would need a buffer, or the data would have to be recorded on a faster medium.

Example 10.6

A data acquisition system is required for an engine test rig. The engines to be tested may have four or six cylinders and may run at any speed up to 6000 rpm. Each cylinder is fitted with a piezoelectric pressure transducer, and the shaft carries a photoelectric, binary position encoder. A set of measurements (one from each cylinder and one from the encoder) is to be taken each time the least significant bit of the encoder changes.

Sketch the layout of a suitable system and compile a table showing the sequence and timing of operations.

Calculate the minimum time required to take one set of measurements using the components specified below. What is the maximum number of sets which could be taken during one revolution? Hence suggest a suitable (binary) number of measurements from each cylinder during one revolution.

How many bits will be needed in the shaft encoder?

What will be the conversion rate of the analogue to digital converter when testing a six-cylinder engine running at full speed? The following specification is provided:

Unit		
Analogue to digital converter:	resolution	= 10 bits
	conversion time	= 20 μs
	input impedance	= 3 kΩ
Sample and hold:	aperture time	= 50 ns
	acquisition time	= 4 μs to 0.1% or 5 μs to 0.01%
Multiplexer:	Settling time	= 200 ns to 0.1% or 250 ns to 0.01%
	ON resistance of switches	= 300 Ω

Solution

The layout of the system is shown in Figure 10.10. A buffer is needed between the multiplexer and the A/D because the input impedance of the A/D (3 kΩ) is not much greater than the ON resistance of the multiplexer switches (300 Ω).This table assumes that data remains available at the output of the A/D until the completion of the next conversion, and that this time will be sufficient for the computer to read and store the data.

From Table 10.8, the time taken for one set of readings is 125.05 μs. The time for one revolution at maximum speed is 60/6000 s, i.e. 10 ms. Therefore, the maximum number of readings which may be taken during one revolution is

$$10\,000/125.05 = 80$$

One set of readings is taken each time the LSB of the encoder changes, so the number of sets taken in a revolution must be a binary number less than 80, i.e. 64. This would require a 6-bit encoder:

$$\text{throughput} = 64 \times 100 \times 6 = 38\,400 \text{ conversions per second}$$

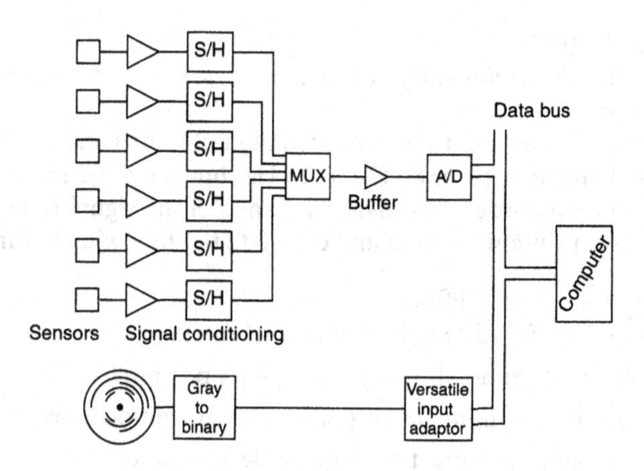

Figure 10.10 *Example 10.6.*

Table 10.8 *Data acquisition sequence and timing for Exercise 10.6*

S/H	*MUX*	*A/D*	*Computer*	*Longest time*	*Time* (μs)
To hold	On ch. 1		read encoder	aperture	0.05
		Convert 1		convert	20
	to ch. 2		read A/D 1	MUX settle	0.2
		Convert 2		convert	20
	to ch. 3		read A/D 2	MUX settle	0.2
		Convert 3		convert	20
	to ch. 4		read A/D 3	MUX settle	0.2
		Convert 4		convert	20
	to ch. 5		read A/D 4	MUX settle	0.2
		Convert 5		convert	20
	to ch. 6		read A/D 5	MUX settle	0.2
		Convert 6		convert	20
Sample	to ch. 1		read A/D 6	acquisition	4
				Total	125.05

Example 10.7
These are the preliminary calculations before starting a project on speech analysis.

Speech signals are to be recorded digitally. The bandwidth of the original signal is from 80 Hz to 15 Hz, but only 100 Hz to 4 kHz need to be recorded. The dynamic range of the signal is 44 dB.

Sketch a suitable system and calculate the following parameters:

1. sampling frequency;
2. converter word length (number of bits);
3. conversion time allowing 8 μs + 8 μs per bit;
4. ratio of minimum signal (rms) to quantization noise (rms);
5. maximum aperture time of sample and hold;
6. maximum acquisition time of sample and hold;
7. droop rate of sample and hold assuming the full range of A/D is 10 V.

The answers to items 4–6 should allow 1 least significant bit error under the worst conditions.

Solution
A sketch of the system is provided by Figure 10.11.

1. The maximum signal frequency is 4 kHz. As the input is a speech waveform, the highest fundamental frequency is unlikely to exceed 400 Hz (440 Hz is the note ‘A’ to which orchestras tune). The higher frequencies are harmonics which carry vital information to distinguish one vowel from another, but which have smaller amplitudes than the fundamentals. According to Shannon’s sampling theorem, the minimum sampling frequency is 8 kHz. In practice a higher sampling frequency should be used to allow for the separation of the pass and stop bands of the anti-aliasing filters. The question does not ask for the design of the filters, so we will suggest a sampling frequency of 12 kHz and filters which pass 4 kHz and stop 8 kHz.
2. The dynamic range is 44 dB, so the ratio of the largest to the smallest signal amplitudes is $10^{44/20} = 158$.

   The next larger power of 2 is 256, i.e. 2^8. Therefore, at least 8 bits are required. For accurate digitization of the quietest speech, more bits would be preferred.

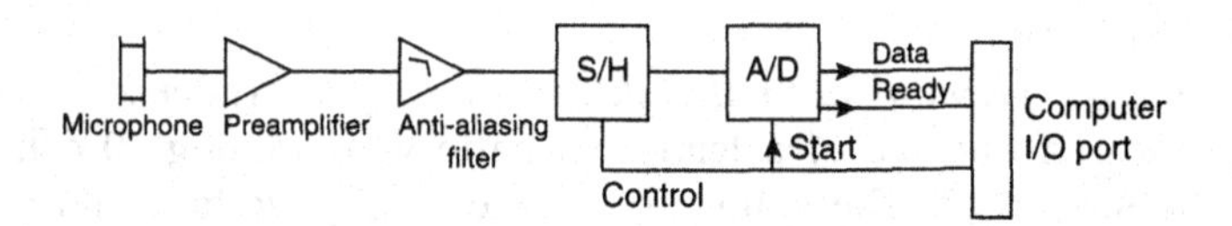

Figure 10.11 *Example 10.8.*

3. Conversion time is $8 + 8 \times 8 = 72$ μs.
4. RMS quantization noise

$$
\begin{aligned}
\text{LSB}/\sqrt{12} &= 0.289 \text{ LSB} \\
\text{maximum signal} &= 256 \text{ LSB peak to peak} \\
\text{minimum signal} &= 256/158 \text{ LSB peak to peak} \\
&= 256/(158 \times 2\sqrt{2}) \text{ LSB rms} \\
\text{Signal-to-noise ratio} &= (256 \times \sqrt{12})/(158 \times 2\sqrt{2}) \\
&= 1.98
\end{aligned}
$$

5. Maximum rate of change of maximum frequency signal is

$$
\begin{aligned}
2\pi \times f_{max} \times V_{max} &= 2\pi \times 4000 \times 2^7 \\
&= 3\,216\,991 \text{ LSB s}^{-1}
\end{aligned}
$$

The signal should not change more than 1 LSB during the aperture time, so the maximum aperture time is $(3.216\,991)^{-1} = 0.31$ μs.

6. The aperture time + conversion time + acquisition time should not exceed the sampling interval:

$$t_{AQ} \leqslant t_{SAM} - t_{CON} - t_{AP}$$

Therefore,

$$t_{AQ} \leqslant \frac{10^6}{12\,000} - 72 - 0.31 = 11 \ \mu\text{s}$$

7. The signal should not droop by more than 1 LSB during the conversion. Therefore,

$$\text{droop} \leqslant \frac{10 \times 10^6}{256 \times 72} = 543 \text{ V s}^{-1}$$

Example 10.8
Unlike the previous examples, this is a low-speed system.

The performance of a lead–acid battery containing 10 cells is to be monitored. The voltage of each cell, total voltage and total current are to be recorded every 5 minutes. A 50 mV shunt is available for measuring the current. The accuracy required is 1 mV in the cell voltage and in proportion for the other measurements.

Design a suitable system.

Solution
There are 10 lead–acid cells. Therefore the battery voltage is a little over 20 V. The required accuracy is 1 mV in 2 V, i.e. 1 in 2000 or 0.05%.

The resolution must be as good as, or better than, the required accuracy, and 1 in 2000 requires 11 bits (1 in 2048) or four digits.

An integrating converter or a dvm would be fast enough since the reading rate is only 1 set in 5 min or 12 readings in 25 s.

A 3½-digit dvm normally has a maximum reading of 1999 and would therefore display only three digits of an input slightly over 2 V.

A 4½-digit dvm would display four digits and resolve to 1 mV on the 20 V range, and in proportion on the 200 V range. Also, a 4½-digit dvm on the 200 mV range would resolve to 10 μV.

A 4½-digit meter should therefore be used. A true four-and-a-half meter, reading to 39 999, would be preferable, but these are uncommon; most 4½-digit meters have a maximum reading of 19 999.

Reed switches would be sufficiently fast and would be necessary for resolving to microvolts when reading the shunt voltage to determine the current. A suitable configuration is shown in Figure 10.12.

When cell no. 7 is being measured, there will be a common mode voltage of 18 V at the LO terminal of the dvm. For 1 mV accuracy the CMRR must not be less than

$$20 \log \frac{18}{0.001} = 20 \log 18\,000 = 85.1 \text{ dB}$$

Measurements are required for an indefinite period so a printer would be a suitable recording device. Whether any calculations are required is not stated in the question, so we do not know

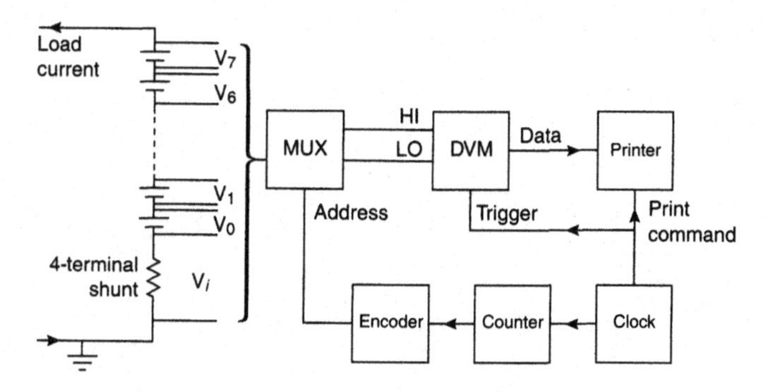

Figure 10.12 *Example 10.9.*

whether a computer will be needed as well. If a computer is included, it could also control the multiplexer switching and trigger the dvm readings at the appropriate times. It could also be used to calculate the power drawn from the battery and monitor how soon each cell was exhausted.

CHAPTER 11

Bus systems

11.1 Introduction

When more than two devices are to be connected together so that any one can communicate with any other, the connections are termed a **bus**. Obviously there must be a certain amount of standardization for all the devices to be compatible, and a large number of standard specifications have been issued by various organizations:

- Institute of Electrical and Electronic Engineers (IEEE);
- Engineering Industries Association (EIA);
- International Electrotechnical Commission (IEC);
- British Standards Institution (BSI).

These specifications are often called just 'standards', but this usage of the word should not be confused with standards of voltage, resistance and so on. The bus systems which are relevant to data acquisition can be broadly classified into busses for microprocessors, for boards and for instruments.

11.1.1 Three-state logic and open collectors

The output pins of two ordinary logic gates cannot be connected together directly because of the confusion which would result if one tried to be high and the other low. Devices connected to a bus must use either three-state logic or open collectors. The three states are output voltage high, output voltage low and high impedance. All the devices connected to the bus are organized so that only one device is active at a time and all the others are in the high impedance state and have no effect on the bus. Each three-state chip has a **chip select** (CS) pin which is usually active-low, and all the outputs are in the high impedance state until the chip is selected. In addition, a bidirectional bus driver chip has **read enable** (RE) and **write enable** (WE) pins to control the direction of data flow. A bar over the top of the symbol CS, for example, indicates 'active low', that is the chip is selected when the pin voltage is low.

Open collector gates have an external collector resistor. If all the gates on the bus share the resistor, the bus goes low if the collector of any gate draws current. In positive logic this is **wired AND** and in negative logic it is **wired OR**.

11.2 Microprocessor busses

A microprocessor communicates with its memory and interface devices by three busses: the data bus (8, 16 or 32 bits wide), the address bus and the control bus. The number of bits in the address bus determines how many memory addresses may be directly addressed. A 16-bit bus can address $2^{16} = 65\,536$ locations, and this is commonly used with an 8-bit data bus. A larger memory can be used by employing indirect addressing methods. The control bus varies from one family of microprocessors to another but they all include signals such 'clock', 'memory address is valid' (VMA or ALE), 'interrupt request' (IRQ or INTR), 'read/write' (R/W), 'reset' and 'halt'.

11.2.1 Interface adapters

Each microprocessor family includes among its support chips an interface adapter. The 6522 versatile interface adapter (VIA) is typical. The VIA requires 16 address locations. It is allocated a base address, say 8000 hexadecimal, which is set in the address decoder. When the processor addresses any location in the range 8000–800F, the chip select input is held low by the address decoder and the chip becomes active. Its 16 registers are selected by the four least significant address lines of the microprocessor. On the output side it has:

- Two 8-bit, bidirectional ports whose directions can be selected individually.
- Four status and control lines which can be used to start conversion, signal conversion complete etc., and which can initiate interrupts.
- Two 16-bit counter timers which can count or generate pulses.

To select which of the many functions the chip is to perform, it must first be set up by writing the appropriate commands into the registers as follows:

- the data direction register (for each port);
- the peripheral control register (for status and control lines);
- the auxiliary control register (to set up the timers).

Data can then be written into or read from the data or the timer registers.

To connect an A/D converter to a microprocessor using an adapter, the data output lines of the A/D are connected to one or more ports programmed to input, the 'convert' input is connected to a 'control' output and the 'ready/busy' output to a 'status' input. The same adapter may also be used to control the multiplexer address and the sample and hold circuit(s).

11.2.2 PC-XT bus

The ubiquitous IBM-compatible PC has slots for extension boards, which were originally intended to drive office machinery. Now it is possible to buy a wide range of data acquisition boards which plug into these slots and, given the right software, turn the computer into a complete measuring instrument.

The processor in the original PC design was an Intel 8-bit machine with separate memory allocated for input and output, in addition to the main memory. Data input and output were controlled by separate I/O read and write lines, whereas data transfer to the main memory was controlled by the memory read and write lines, and the extension sockets were fitted with 62-pin edge connectors whose pin connections are shown in Table 11.1. This is the XT bus otherwise known as the industry standard (ISA) 8-bit bus. Not all the signals on the bus are available for extension cards; as the same bus is carrying signals to the hard and floppy discs. Interrupt request 2 and its acknowledgement are reserved for users.

11.2.3 PC-AT bus

When the 16-bit 286 processor came on the scene, the AT bus was introduced, but so that boards designed for the XT bus could still be used, the XT bus was not replaced, but extended, by the

Table 11.1 *The industry standard bus for PCs*

Description	*Solder side*	*Pin no.*	*Pin no.*	*Component side*	*Description*
	GND	B1	A1	–I/O CHCK	Error check
From motherboard	RESET DRV	B2	A2	D7	Data I/O 7
	+ 5 V	B3	A3	D6	
Interrupt request 2	IRQ2	B4	A4	D5	
	–5 V	B5	A5	D4	
	DRQ2	B6	A6	D3	
	–12 V	B7	A7	D2	
	–CARDSLCTD	B8	A8	D1	
	+ 12 V	B9	A9	D0	Data I/O 0
	GND	B10	A10	I/O CHRDY	
Memory write	–MEMW	B11	A11	AEN	Address enable
Memory read	–MEMR	B12	A12	A19	Address 19
I/O write	–IOW	B13	A13	A18	
I/O read	–IOR	B14	A14	A17	
	–DACK3	B15	A15	A16	
	DRQ3	B16	A16	A15	
DMA acknowledge 1	–DACK1	B17	A17	A14	
DMA request 1	DRQ1	B18	A18	A13	
DMA acknowledge 0	–DACK0	B19	A19	A12	
Clock	CLK	B20	A20	A11	
Interrupt request 7	IRQ7	B21	A21	A10	
	IRQ6	B22	A22	A9	
	IRQ5	B23	A23	A8	
	IRQ4	B24	A24	A7	
	IRQ3	B25	A25	A6	
	–DACK2	B26	A26	A5	
	T/C	B27	A27	A4	
Address logic enable	ALE	B28	A28	A3	
	+ 5 V	B29	A29	A2	
	OSC	B30	A30	A1	
	GND	B31	A31	A0	Address 0

addition of another 36-pin connector carrying the extra eight data and address lines and other control lines (Fairhead, 1992). This is the ISA 16-bit bus, which has also been used with 386 and 486 processors.

11.2.4 VESA and EISA local busses

In 1992, the Video Electronics Standards Association (VESA) proposed an extension to the 16-bit ISA bus, called the VL bus, which has been widely advertised as the VESA bus.

A third socket has been added to extend the bus to 32 bits to work with machines based on the 486 processor, while still remaining compatible with extension boards designed for earlier machines. This is the EISA bus (Gann, 1994).

11.2.5 PCI bus

The quest for faster and faster transfer of more and more data has led to the development of the peripheral component interconnect (PCI) bus. Although it was first proposed by the microprocessor manufacturer Intel, the PCI bus is independent of the type of processor. The current version transfers data in 32-bit words and is clocked at 33 MHz. A 64-bit version is planned. Boards designed for the earlier busses may still be used in the slots of a PCI bus. Another feature of this bus is that the various settings which are necessary before a board can function like input/output addresses and interrupt request settings can be set entirely in software. This is the widely advertised 'plug and play', and the software is incorporated in Microsoft Windows 95. On the older busses, it was necessary to set switches or jumpers on the board and match the software to them. More details are given by Moralee (1995) and Wright (1995).

11.2.6 PCMCIA bus for laptop computers

An interface board for a laptop computer must be small and consume minimal power. The Personal Computer Memory Card International Association standard specifies a 27-pin interface connector and cards 85.6 × 54 mm and 3.3, 5 or 10 mm thick. As the name implies, this system was originally developed for memory modules, but now a growing range of cards for both digital and analogue inputs is available. It is standard practice when inserting or removing a connector to switch off the power first, but the PCM-CIA is designed so that cards can be changed with the power on.

11.3 Busses to link boards

Many instruments use a combination of bought-in and purpose-built boards. A number of manufacturers sell a selection of boards from which the user can build a data acquisition system, and there are now a number of standard bus systems for this purpose. Using a standard bus system allows the user to mix boards from different manufacturers and plug them into a common backplane. STE, VME, Multibus and Futurebus were reviewed by Edwards (1987).

11.3.1 STD and STE

The STD (IEEE 961; 1978) and its European derivative STE (IEEE P1000; 1987) were designed for 8-bit microprocessor systems. They use 100 × 160 mm 'Eurocards', STD uses 56-way edge connectors, but STE uses DIN 41612 (IEC 603-2) connectors which have two rows of 32 pins each. These carry power at + 5 V and 12 V, have eight data lines, 16 or 20 address lines, and 22 control lines. The maximum data transfer rate is five megabytes per second. All the rules governing data transfer are laid down in the IEEE standards, and further details are given by Dettmer (1987).

11.3.2 VME

The VME bus (IEEE 1014) was introduced by Motorola, but it can be used with any company's products. It was originally developed for 16-bit microprocessors, but can now be used with 16 or 32 bits. It uses double Eurocards, and two DIN 41612 connectors to each board. Each connector has three rows of 32 pins. The data transfer rate can be at least 20 megabytes per second.

11.3.3 Futurebus

Futurebus (IEEE 896.1) is designed for high-speed, asynchronous data transfers with a 32-bit data bus. It uses triple Eurocards, and a single 5 V supply. The name comes from the designers' intention to provide for future developments in technology (Borrill, 1987).

11.4 Asynchronous serial data transmission

The simplest bus systems transmit data serially, that is one bit after another, so only two wires are required to carry the data, although others may be used for control signals. They are descendants of the old telegraph systems.

Both transmitter and receiver have separate clocks running at approximately the same frequencies. Each byte of data is encoded in ASCII code and transmitted down a pair of wires as a sequence of logic levels. Each byte is preceded by a low-level 'start' bit and followed by two high-level 'stop' bits. Thus, 11 bits are necessary to transfer each byte of data. The rate of data transfer in bits per second is called the **baud** rate.

Encoded in this format, data may be transmitted over a pair of dedicated wires, or over the public telephone lines using a **modem** (modulator–demodulator) at each end. Modems are also known as datasets or data communication equipment (DCE). **Duplex** transmission means that data may be transmitted in both directions simultaneously, and **half duplex** means that it may be transmitted in either direction, but only one at a time.

The standard connector (which is not always used) is a 25-pin, subminiature D-type with the female connector on the data terminal equipment (DTE), such as the computer, and the male on the modem. However, IBM PCs and clones use a 9-pin subminiature D-type.

The essential connections are 'transmit and receive data' and 'signal ground'. If data can be transmitted along the bus in either

Table 11.2 *Connections for RS-232 interface*

Function	*Mnemonic*	*Pin number*	
		25 pin	*9 pin*
Protective ground	GND	1	–
Transmitted data	TXD	2	2
Received data	RXD	3	3
Signal ground (common return)		7	5
Request to send	RTS	4	7
Clear to send	CTS	5	8
Data set ready	DSR	6	6
Data terminal ready	DTR	20	4

direction, then handshaking signals are also necessary to control the flow of data. These are 'request to send' (RTS) and its reply 'clear to send' (CTS), and 'data terminal ready' and its reply 'data set ready'. Table 11.2 shows the pin allocations for 25- and 9-pin connectors. Note that if two DTEs, such as two computers, are connected together, the 'transmit' pin for one must be connected to the 'receive' pin for the other, and vice versa. Similarly, RTS is connected to CTS and vice versa.

11.4.1 Standards

The various standards for serial data transmission differ in their logic levels.

Current loop

Logic 1 is represented by a current of 20 mA, and logic 0 by 4 mA. No current at all indicates a fault.

EIA RS-232-E (1992); European equivalent V28

Open circuit voltages are logic 0 = + 3 to + 25 V and logic 1 = –3 to –25 V. When the lines are terminated, control signals ON or logic 0 = + 5 to + 15 V and control signals OFF or logic 1 = –5 to –15 V.

The maximum load is 2500 pF (the older RS-232-C specified that maximum transmission distance should be 50 ft), and the maximum transmission rate is 20 kbaud.

EIA RS-423

Open circuit voltages are logic 0 = + 0.2 to + 12 V and logic 1 = –0.2 to –12 V. When the lines are terminated, logic 0 = + 3.6 to + 5.4 V and logic 1 = –3.6 to –5.4 V. The maximum transmission distance is 2000 ft and the transmission rate is 300 kbaud.

EIA RS-422

This is similar to RS-423, but differential, which improves the common-mode rejection.

Open circuit voltages are logic 0 = + 0.2 to + 7 V and logic 1 = –0.2 to –7 V. When the lines are terminated, logic 0 = + 2 to

Table 11.3 *Universal asynchronous receiver–transmitter*

Section	*Function*	*Inputs*	*Outputs*
Receiver	detects start and stop bits, checks parity	serial data, clock, register disable	parallel data
Transmitter	adds start, stop and parity bits	parallel data, clock, register load	serial data
Control	controls resets, word length and parity, indicates empty registers and errors	control	status, error flags

+ 5 V and logic 1 = –2 to –5 V. The maximum transmission distance is 4000 ft, and the transmission rate 10 Mbaud.

EIA RS-485

Like RS-422, this bus operates on differential signals. Three-state drivers are used for both receivers and transmitters, so that up to 32 drivers and 32 transmitters may be connected to one bus. The maximum speed and transmission distance are the same as RS-422, but the operating voltage is higher.

11.4.2 Asynchronous interface transceivers

Parallel data can be interfaced to and from a serial bus by means of an integrated circuit known as a universal asynchronous receiver transmitter (UART) such as the 6402 (Table 11.3). For further details refer to manufacturers' data sheets or Zaks (1977).

11.5 General purpose interface bus

11.5.1 Origins

For many years the choice of instruments which could be linked together was limited by the variety of connectors, voltage levels, digital codes and control signals. The advent of computers, cheap enough to be used as part of an instrumentation system, made the need for standardization more urgent. Just as a telephone is quite useless unless the people you wish to talk to also have compatible telephones, an instrument with digital output is similarly useless unless compatible controllers and recorders exist.

The US instrument and computer company Hewlett-Packard makes such a wide range of instruments that they were able to develop an interfacing standard and market complete systems. This was marketed as HP-IB. Once it was introduced, it gained wide acceptance, and instruments to work with this bus are available from all the major manufacturers (Knoblock *et al.*, 1975).

The specification for the bus was issued by the Institution of Electrical and Electronic Engineers in 1978 and updated in 1987 to IEEE STD 488.1-1987. The corresponding British standard is BS 6146:1981. These are almost identical with IEC 625-1. The only difference is that the IEEE bus uses a 24-pin connector and IEC specifies a different, 25-pin, connector. The bus is now usually known as the general purpose interface bus (GPIB), the IEEE-488 bus, or even just the IEEE bus. The introduction of the first IEEE standard 488 in 1975 made a big step forward in computer-controlled measurement, but the content of the bus messages was left to individual instrument designers until IEEE-488.1 was revised and IEEE-488.2 introduced in 1987. The software was further standardized by the introduction in 1990 of a new programming language known as SCPI (IOtech, 1991).

The full titles of the current standards are:

- IEEE-488.1 – 1987, IEEE Standard (Reaffirmed 1994) Digital Interface for Programmable Instrumentation;
- IEEE-488.2 – 1992, Codes, Formats, Protocols, and Common Commands for use with IEEE-488.1-1987;
- SCPI – Standard Commands for Programmable Instrumentation.

The description below should give a good idea of the basic operation of the GPIB bus, but the standards include further refinements which are not included here. More information can be found in the standards documents and in the references.

11.5.2 Device classification

Whole books have been written about the GPIB bus (e.g. Caristi, 1989), so what follows is only a summary. The system works on the principle of a number of people (i.e. instruments) having a meeting. At any given time, there can be only one chairman, and

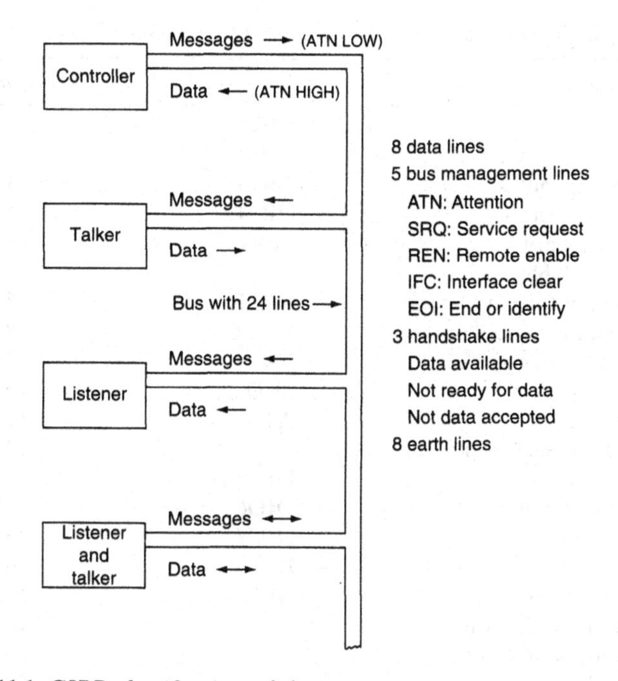

Figure 11.1 *GIPB classification of devices.*

only one talker, but any number of listeners. Usually, but not always, the chairman will not change during the meeting, but he will call on many members to speak. All devices connected to the bus are classified as controllers, listeners or talkers (Figure 11.1):

- The controller corresponds to the chairman of the meeting, and may be either a computer or a microprocessor fitted with a suitable interface.
- A talker is any source of data such as a measuring instrument.
- A listener is any device which receives data, such as a printer.

An instrument such as digital voltmeter may be a listener while the controller sets its range and mode, and then becomes a talker to transmit the result of the measurement to any device which has been programmed to listen. The minimum system comprises one talker, one listener and no controller, but this is not an economic arrangement.

Table 11.4 *Pin connectors for the GPIB standard connector*

Function	*Mnemonic*	*Pin no.*
Data in or out 1	DIO1	1
Data in or out 2	DIO2	2
Data in or out 3	DIO3	3
Data in or out 4	DIO4	4
End or identify	EOI	5
Data available	DAV	6
Not ready for data	NRFD	7
Not data accepted	NDAC	8
Interface clear	IFC	9
Service request	SRQ	10
Attention	ATN	11
Shield (connect to ground)		12
Data in or out 5	DIO5	13
Data in or out 6	DIO6	14
Data in or out 7	DIO7	15
Data in or out 8	DIO8	16
Remote enable	REN	17
Ground, twisted with 6		18
Ground, twisted with 7		19
Ground, twisted with 8		20
Ground, twisted with 9		21
Ground, twisted with 10		22
Ground, twisted with 11		23
Signal ground		24

11.5.3 Bus lines

There are 24 bus lines, and each device is connected to all of them (Figure 11.1 and Table 11.4).

The standard specifies TTL levels and open collector drivers for the handshake lines and open collector or three-state drivers for the data and bus management lines. Negative logic is used, thus 1 = true = low and 0 = false = high.

11.5.4 Data lines

Data is transmitted byte serial, bit parallel, that is eight bits at a time. Each 8-bit byte usually represents one character according to ASCII code, so that an instrument can return an alphanumeric message such as 123.4 V. The most significant character

Table 11.5 *ASCII and GPIB code chart*

BITS B7 B6 B5 / B4 B3 B2 B1	Ø Ø Ø	Ø Ø 1	Ø 1 Ø	Ø 1 1	1 Ø Ø	1 Ø 1	1 1 Ø	1 1 1
	CONTROL		NUMBERS SYMBOLS		UPPER CASE		LOWER CASE	
Ø Ø Ø Ø	0 **NUL** 0 0	20 **DLE** 10 16	40 0 **SP** 20 32	60 16 **0** 30 48	100 0 **@** 40 64	120 16 **P** 50 80	140 0 **'** 60 96	160 16 **p** 70 112
Ø Ø Ø 1	1 GTL **SOH** 1 1	21 LLO **DC1** 11 17	41 1 **!** 21 33	61 17 **1** 31 49	101 1 **A** 41 65	121 17 **Q** 51 81	141 1 **a** 61 97	161 17 **q** 71 113
Ø Ø 1 Ø	2 **STX** 2 2	22 **DC2** 12 18	42 2 **"** 22 34	62 18 **2** 32 50	102 2 **B** 42 66	122 18 **R** 52 82	142 2 **b** 62 98	162 18 **r** 72 114
Ø Ø 1 1	3 **ETX** 3 3	23 **DC3** 13 19	43 3 **#** 23 35	63 19 **3** 33 51	103 3 **C** 43 67	123 19 **S** 53 83	143 3 **c** 63 99	163 19 **s** 73 115
Ø 1 Ø Ø	4 SDC **EOT** 4 4	24 DCL **DC4** 14 20	44 4 **$** 24 36	64 20 **4** 34 52	104 4 **D** 44 68	124 20 **T** 54 84	144 4 **d** 64 100	164 20 **t** 74 116
Ø 1 Ø 1	5 PPC **ENQ** 5 5	25 PPU **NAK** 15 21	45 5 **%** 25 37	65 21 **5** 35 53	105 5 **E** 45 69	125 21 **U** 55 85	145 5 **e** 65 101	165 21 **u** 75 117
Ø 1 1 Ø	6 **ACK** 6 6	26 **SYN** 16 22	46 6 **&** 26 38	66 22 **6** 36 54	106 6 **F** 46 70	126 22 **V** 56 86	146 6 **f** 66 102	166 22 **v** 76 118
Ø 1 1 1	7 **BEL** 7 7	27 **ETB** 17 23	47 7 **'** 27 39	67 23 **7** 37 55	107 7 **G** 47 71	127 23 **W** 57 87	147 7 **g** 67 103	167 23 **w** 77 119
1 Ø Ø Ø	10 GET **BS** 8 8	30 SPE **CAN** 18 24	50 8 **(** 28 40	70 24 **8** 38 56	110 8 **H** 48 72	130 24 **X** 58 88	150 8 **h** 68 104	170 24 **x** 78 120
1 Ø Ø 1	11 TCT **HT** 9 9	31 SPD **EM** 19 25	51 9 **)** 29 41	71 25 **9** 39 57	111 9 **I** 49 73	131 25 **Y** 59 89	151 9 **i** 69 105	171 25 **y** 79 121
1 Ø 1 Ø	12 **LF** A 10	32 **SUB** 1A 26	52 10 ***** 2A 42	72 26 **:** 3A 58	112 10 **J** 4A 74	132 26 **Z** 5A 90	152 10 **j** 6A 106	172 26 **z** 7A 122
1 Ø 1 1	13 **VT** B 11	33 **ESC** 1B 27	53 11 **+** 2B 43	73 27 **;** 3B 59	113 11 **K** 4B 75	133 27 **[** 5B 91	153 11 **k** 6B 107	173 27 **{** 7B 123
1 1 Ø Ø	14 **FF** C 12	34 **FS** 1C 28	54 12 **,** 2C 44	74 28 **<** 3C 60	114 12 **L** 4C 76	134 28 **** 5C 92	154 12 **l** 6C 108	174 28 **¦*** 7C 124
1 1 Ø 1	15 **CR** D 13	35 **GS** 1D 29	55 13 **-** 2D 45	75 29 **=** 3D 61	115 13 **M** 4D 77	135 29 **]** 5D 93	155 13 **m** 6D 109	175 29 **}** 7D 125
1 1 1 Ø	16 **SO** E 14	36 **RS** 1E 30	56 14 **.** 2E 46	76 30 **>** 3E 62	116 14 **N** 4E 78	136 30 **Λ** 5E 94	140 14 **n** 6E 110	176 30 **~** 7E 126
1 1 1 1	17 **SI** F 15	37 **US** 1F 31	57 15 **/** 2F 47	76 UNL **?** 3F 63	117 15 **O** 4F 79	137 UNT **_** 5F 95	157 15 **o** 6F 111	177 **DEL (RUBOUT)** 7F 127
	ADDRESSED COMMANDS	UNIVERSAL COMMANDS	LISTEN ADDRESSES		TALK ADDRESSES		SECONDARY ADDRESSES OR COMMANDS (PPE)	(PPD)

*¦ ON SOME KEYBOARDS OR SYSTEMS

KEY

octal	25 PPU	GPIB code
	NAK	ASCII character
hex	15 21	decimal

is transmitted first. (For higher data transfer rates each byte can be programmed to encode two decimal or hexadecimal digits in binary code.) The end of the data is signalled by making the EOI line true while the last character is being transmitted. (The 1978 edition of the standard allowed the use of CR and LF characters as terminators, but this is a possible source of confusion if the controller is expecting the EOI line to be used.)

Table 11.5 shows how the eight bits are interpreted, and the key to this chart is at the bottom left of the table. Each square represents a 7-bit binary number with the three most significant bits shown in the column headings and the four least significant bits in the rows. The **heavy type** in the centre of each square is the ASCII character. The value in hexadecimal code is the small number in the bottom left of each square and the decimal value is in the bottom right. For example, 001 0101 in binary code is 15 in hexadecimal. It is also 21 in decimal or ASCII character NAK (negative acknowledge).

When the ATN line is low (true), the character represents a bus message and the meaning is the small figure in the top right. Thus, when the ATN line is low, 001 0101 can also represent the GPIB command PPU (parallel poll unconfigure). Similarly, when ATN is low, 011 0100 represents the message 'listen address = 20 decimal', but when it is high (false) the same code represents the digit '4'. The data lines are also used to carry commands, some of which are illustrated in Table 11.6.

11.5.5 Bus management lines

The direction and route of data flow is controlled by these lines. This is illustrated in Table 11.6 and below.

ATN: attention

This line is driven by the controller and functions like the shift key. When ATN is true (low), the signals on the data lines are interpreted as bus messages, but when ATN is false (high), they are alphanumeric data. Thus, when ATN is low, 0011 0100 represents the message 'listen address = 20 decimal', but when ATN is high (false), it represents the digit '4'.

IFC: interface clear

This line is driven by the controller and is pulsed low to set all devices on the bus to their idle state. It overrides all other activity.

REN: remote enable

This line is driven by the controller. When an instrument receives a low on this line, the front panel controls are disabled and the instrument is controlled by bus messages from the controller unit until a 'go to local' (GTL) message is sent along the data lines.

SRQ: service request

This line is driven by either listener or talker to ask for service from the controller, such as the 'end of conversion' message from a dvm or 'out of paper' message from a printer. To use service requests, the program must initiate a serial poll or parallel poll to identify the source of the request whenever the controller detects SRQ. In our analogy of a meeting, a service request corresponds to a member of the meeting raising a hand to attract the chairman's attention.

EOI: end or identify

This line serves two entirely different functions. It may be driven by a talker to mean 'end of this data string' or it may be combined with the ATN line true (low) to initiate a parallel poll, and identify the source of an interrupt.

11.5.6 Device addressing

Before data can be transferred, the appropriate devices must be set to talk and to listen. This is done by the controller. Each device on the bus is allocated a 5-bit **primary address**. Hence 31 devices may be addressed. Hexadecimal codes 20 to 3E represent the 31 listen addresses from 0 to 30 decimal: 3F is UNL (unlisten) which returns the previous listener(s) to the idle state; similarly 40–5E represent talk addresses and 5F is 'untalk'.

The user sets the addresses of the devices in the system to correspond to those in the program by means of small switches which are usually located on the rear panel near to the IEEE connector, but they may be inside the case. Some instruments also use a **secondary address** to differentiate between different functions within the same instrument. Up to 15 instruments may be connected at once.

An example of a simple application is given in Example 11.2.

11.5.7 Interrupts

Any device may request the attention of the controller by asserting the SRQ line, i.e. pulling it low. When the controller detects this, it must find out which device initiated the request and why. There are two ways it can do this, by parallel or serial poll. Parallel is faster, and is limited to eight devices at any one time, but serial supplies more information. Each device has a status byte and, for a parallel poll, each device asserts one bit. When the controller detects an interrupt it asserts the ATN and EOI lines together, thus saying 'identify yourselves'. The devices then put their status bytes on the data lines, and whichever byte is low indicates a device requesting service. Obviously before this can be used, the controller must know which bit represents which device, so this is set up initially using the addressed command 'parallel poll config' (PPC), hexadecimal 5. The configuration can be changed if necessary using the unaddressed command 'parallel poll unconfig' (PPU), hexadecimal 15. A serial poll interrogates each device in turn. The controller first sends the unaddressed command 'serial poll enable' (SPE), hexadecimal 18, and then addresses each device in turn. As it is addressed, the device puts its status byte on the data lines. The meaning of bits within the status byte used to be left to the instrument designer, but in IEEE 488-2 it is standardized. Having identified the cause of the interrupt, the controller can return the system to normal using 'serial poll disable' (SPD), hexadecimal 19.

11.5.8 Handshake

The three-line handshake controls the transfer of each byte from the talker to the data lines and from the data lines to the listener(s).

NRFD: not ready for data

This line is driven by the listener(s): low = not ready; high = ready. The use of open collector outputs ensures that this line stays low until all listeners are ready.

NDAC: not data accepted

This line stays low until all listeners have accepted the data; it is driven by the listener(s).

DAV: data available

This line is driven by the talker.

The handshake is illustrated in Figure 11.2. In the initial state no data is available, so DAV is false (high); data has not been accepted so NDAC is true (low), and the listeners are not ready for data so NRFD is also true.

When a listener has been addressed and is ready to receive data, it tries to pull the NRFD line false. Because this line is connected to open collector drivers sharing a common load, the line will not go false until all addressed listeners are ready. Unaddressed listeners have no effect. When NRFD goes false the talker replies 'now you are all listening, I will tell you something'. In electronic terms it sets the DAV line true, and puts its data on the data lines. Each listener replies by pulling NRFD true again, thus saying 'one word at a time, please'. It then reads the data lines and effectively says 'got that one, thank you' by trying to pull the NDAC line false. The line does not actually go false until all the listeners have accepted the data. The talker replies by making the DAV line false again, and then it can safely change the data lines. The return of the DAV line to false makes the listeners return the NDAC line to true, thus restoring the initial conditions. If another byte of data is to be transmitted, the process is repeated but, during the last byte, the EOI line is made true. Thus when the ATN line is false (high), asserting the EOI line means 'end of message'.

11.5.9 Interface boards and chips

To use the GPIB bus with a computer, interface circuits are needed. These include transceivers to take data from and return it to the computer's data and address busses, and logic circuits to control handshaking and interrupt handling. Other transceivers, which must be correctly terminated, are required as drivers for the bus lines. Interface boards which plug into the expansion slots of PCs or other computers are made by several manufacturers, and the software to drive the bus from one or more of the popular high-level languages is sold as a package with the boards.

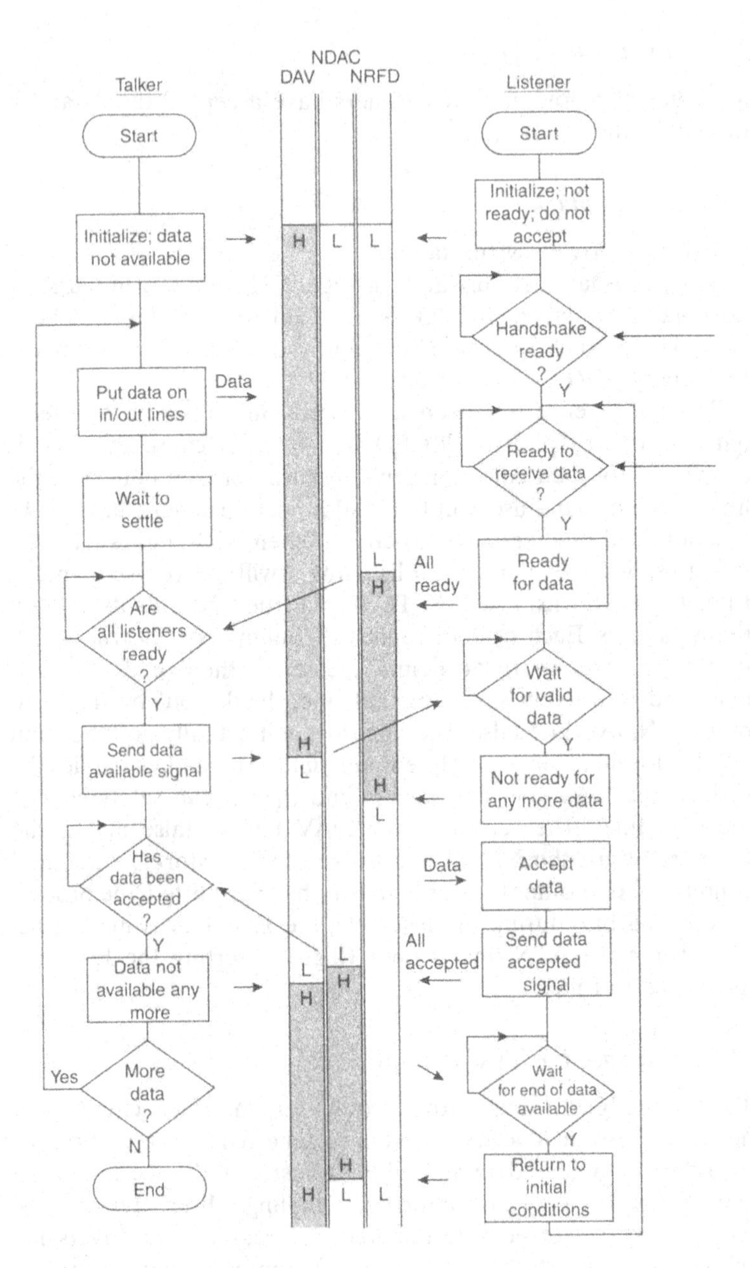

Figure 11.2 *GPIB handshake.*

The logic circuitry necessary to drive the IEEE bus has been simplified by the introduction of dedicated ICs for each family of microprocessors:

- Motorola transceiver MC3447, interface MC68488;
- Intel transceiver P8291, interface P8293;
- NSC transceivers DS75160 for the data bus, and 75161 or 75162 for the management bus;
- NEC intelligent interface controller μPD7210;
- Texas Instruments bus transceiver with open collector driver MC3446N and GPIB adaptor 9914A;

A more complex chip designed for instruments is NAT 4882. This is compatible with NEC μPD7210 and TI 9914A.

Converters are also available from RS-232 and RS-422 to GPIB so that a group of GPIB instruments may be controlled by a computer which has only a serial port.

11.5.10 Standard commands for programmable instrumentation

The introduction of the first IEEE standard 488 in 1975 made a big step forward in computer-controlled measurement, but there were still various parameters left to the choice of individual instrument designers, which meant that the software to drive instruments through the bus was far from standard. This led to the introduction of the Tektronix 'Codes and Formats' standard which was incorporated in the new IEEE standard IEEE 488-2 in 1987. At the same time the original standard was updated to become IEEE 488-1. In 1990 a consortium of instrument manufacturers introduced the Standard Commands for Programmable Instrumentation, which became known as SCPI (IOtech, 1991).

SCPI defines the character string which must be sent down the bus to make the instrument perform a specific function, for example to command it to measure d.c. current. It also defines the format of the returned data such as '3.45E-1'. A SCPI command decoder is situated in each instrument. SCPI messages may also be transmitted over RS-232 or VXI bus systems.

11.6 VXI bus

VXI stands for VME with extensions for instruments. It was developed for the US Air Force by a consortium of instrument manufacturers and was introduced in 1987. It is intended for much larger systems than computer extension boards can handle. The system uses a standard card cage with a VME backplane with additional lines, so it can accept both VXI and VME boards. The computer circuit may be on one of the boards (embedded), or it may be free standing and linked to the VXI system by a GPIB bus. Many types of data acquisition board are available in this system. They may be **register-based**, programmed at low level and operating at high speed (up to 40 megabytes per second), or they may be **message-based** and communicate in a similar way to GPIB devices. The use of the GPIB bus simplifies the programming, but reduces the speed of the system.

11.7 Examples

Example 11.1

1. What is meant by describing an A/D converter as microprocessor compatible?
2. How would you connect the 'chip select', 'chip enable', A_0 and 'read/convert' pins on an A/D to an 8-bit microprocessor such as type 6800? The A/D is to be located at addresses 8000 and 8001 hexadecimal.
3. By means of a flow chart, or otherwise, show how you would program a microprocessor to initiate a conversion and read the result back into memory locations hexadecimal 1000 and 1001 after 40 μs.
4. How would you connect the 'ground' pins on the A/D if the input signal was completely floating?
5. If the A/D is to be used over a chip temperature range from 0 °C to 40 °C, and an accuracy of $\pm$ 1 LSB is required, would the internal reference be sufficiently stable?
6. The input was connected to the $\pm$ 5 V range pin, and a conversion gave 0011 0111 at location 1000 hexadecimal and 1101 0000 at location 1001 hexadecimal. What was the input voltage?

Some relevant details from the A/D specification are:

Resolution	12 bits
Reference voltage	10 V
Code	Offset binary

Temperature coefficient using internal reference:

Grade J	50 ppm
Grade K	27 ppm
Grade L	10 ppm

Solution

1. A microprocessor-compatible A/D converter has three-state data outputs and can be connected to the microprocessor data bus directly without an interface.

2. A 12-bit converter must send its data to an 8-bit microprocessor in two bytes, and it is conventional to allocate the 8 MSB of data to the high byte and the 4 LSB of data plus four zeros to the low byte. The lowest bit of the address bus, A_0, is connected directly to the byte address (A_0) pin on the A/D. Thus an even-numbered address reads the high byte and an odd-numbered address reads the low byte and the four trailing zeros. In this example, all lines of the address bus except the LSB are connected to a decoder, arranged so that when either location 8000 or 8001 is addressed, the decoder sends the chip select pin low.

   Phase 2 of the microprocessor clock is connected to chip enable.

   Read/write on the microprocessor is connected to the ready/convert pin on the A/D, so that a write operation to the selected address starts a conversion.

3. Flow chart:

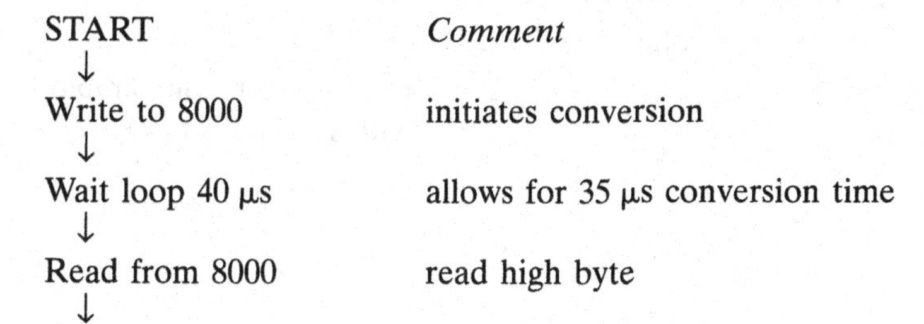

Store result at 1000 store

↓

Read from 8001 read low byte

↓

Store result at 1001 store

↓

END

4. The analogue and digital ground pins on the converter IC should be connected to the ground pins of the analogue and digital supplies, respectively. As the input is floating, the analogue ground should be connected to the digital ground at the supplies.
5. The reference is 10 V and the resolution is 12 bits. Therefore, one least significant bit = $V_{REF}/4096$.

   The allowable change in V_{REF} over working temperature range is

   $$\pm 1 \text{ LSB} = \pm 1 \text{ in } 4096 = \pm 0.000\,244$$

   The allowable change in V_{REF} per degree C, i.e. the temperature coefficient, is

   $$\pm 1 \text{ in } 4096 \times 40 = \pm 0.000\,006\,1 = \pm 6.1 \text{ ppm}$$

   Therefore, a very high stability external reference would have to be used with a temperature coefficient better than 5 ppm.
6. MSB = 0011 0111; LSB = 1101 (four trailing zeros)

   $$\text{digital O/P of A/D} = 0011\ 0111\ 1101$$
   $$= 1 + 4 + 8 + 16 + 32 + 64 + 256 + 512 = 893$$
   $$\text{voltage} = 10 \times 893/4096$$
   $$= 2.18017$$

   In offset binary, this is relative to –5 V. Therefore,

   $$\text{actual voltage} = -5 + 2.18017$$
   $$= -2.820 \text{ V (to resolution comparable with LSB} = 2.4 \text{ mV)}$$

Example 11.2
Compile a table showing the sequence of signals on the ATN, EOI, IFC and data lines to set the range and function of a dvm and read back the result. Show how to set the dvm addresses to 14 decimal on a binary switch.

Solution
Decimal 14 = 000 1110 in binary so, to set the address switch to decimal 14, switch bits 1, 5, 6 and 7 OFF and bits 2, 3 and 4 ON. Table 11.6 shows a simple sequence of signals to set the range and function of a dvm and read the result. In practice this would probably be supplemented by a trigger command (GET = 08 hex) and an interrupt request when the reading is ready.

Table 11.6 *Sequence of signals to set the range and function and then read a dvm whose address is set to 20 decimal = 14 hexadecimal = 10100 binary*

ATN	*EOI*	*IFC*	*Data*	*Mnemonic*	*Action*
L	H	L		IFC	interface clear = initialize
L	H	H	3F	UNL	unlisten
L	H	H	34	MLA 14	device 14 becomes a listener
H	H	H	1st control byte		ASCII codes to set range and function;
H	H	H	2nd control byte		These codes used to be defined
H	H	H	3rd control byte		by the instrument manufacturer,
H	H	H	etc.		are now standardized by SCPI
H	L	H	last control byte		
L	H	H	3F	UNL	unlisten
L	H	H	54	MTA 14	device 14 becomes a talker
H	H	H	most significant data byte		
H	H	H	2nd data byte		ASCII codes sending
H	H	H	etc.		one digit at a time
H	H	H	etc.		
H	L	H	least significant data byte		
L	H	H	5F	UNT	untalk; stops all listeners

CHAPTER 12

Software for data acquisition

12.1 Introduction

This chapter is intended to classify and review the functions commercially available. Examples of each type of program are given, but no attempt is made to recommend particular software because new programs and packages are coming on the market so frequently.

Soon after minicomputers came on the market, they were adapted for data acquisition, and in the 1960s Digital Equipment Corporation's PDP-8 was widely used for this purpose. Fortunately these early computers were sold with very comprehensive technical information, literally down to the last screw, but memory was puny by today's standards. The author's first machine had 4 kilobytes (sic) of RAM and its only nonvolatile memory was paper tape (Taylor, 1974). Compare that with a modern machine with 8 megabytes of RAM and several hundred megabytes of nonvolatile hard disk.

The advent of microcomputers was a major step forward because they were so much cheaper than minis. The Commodore PET was very convenient because it used the GPIB bus as its normal means of communicating with peripherals, and so GPIB instruments could be plugged in without the need to install an interface in the computer. Acorn's 'BBC' computer was designed for teaching and was well documented. It allowed the user access to the microprocessor bus, a serial port and a parallel one known as the 'user port'.

IBM introduced their 'personal computer' originally as a business machine, but provided slots for extension cards with access to the computer bus, the ISA bus in Table 11.1. These were intended for interfaces to other office equipment such as modems, but engineers soon realized that this arrangement could be used for other purposes. Now, many cards are available with data acquisition systems built on them so that, with appropriate cards installed, the analogue signals may be connected to the back of the computer which then collects, processes, displays and records the data. In addition these machines had serial and parallel ports. Although still called PCs, the successors to this design, and their clones, have changed almost beyond recognition. At the same time Apple have developed their range of Macintosh computers. Boards and software are now widely available for both and for laptop computers. The enormous increase in the availability of memory at a reasonable price has permitted the development of programs controlled by the user operating a mouse and pointing to an icon or symbol on the screen. Work is well advanced on voice-controlled commands. These developments make computer-driven systems available to users with no programming experience.

Small data acquisition units are also available which plug into the parallel port of a PC. These units are convenient for laptop computers, but on a PC the parallel port is often required for a printer and is not available for inputs.

12.2 Software choices

Software for a simple data acquisition system will often be written in assembly language by the designer. Many data acquisition systems use ready-made acquisition boards plugged into the expansion bus of a PC, a Macintosh or laptop computer. This gives the user access to much sophisticated software without the need to write any programs at all.

Having designed the system as indicated in Chapter 10, the user selects a board or boards from the very wide and expanding range on the market. The board manufacturers also sell software for many purposes, and there are other companies which specialize in, and advise on, scientific software including programs for chemical, biological and medical uses as well as the more general programs described below.

12.2.1 Criteria for choosing software

To choose from the mass of software available, consider the following points:

1. The computer to be used:

(a) IBM-XT (8-bit) PC or clone, based on 8088 processor;
(b) IBM-AT (16-bit) PC or clone, based on 80286, 386 or 486 processor;
(c) IBM PS/2 with microchannel bus;
(d) Macintosh with NuBus;
(e) Macintosh with other busses, LC or SE;
(f) PowerMac or Pentium PC with PCI bus;
(g) workstation;
(h) laptop with PCMCIA bus.

Most of the programs described below require at least a 16-bit machine.

2. The operating system – the choice lies between:
   (a) DOS;
   (b) Windows 3.1 or Windows 95;
   (c) PowerMac or Mac™ OS;
   (d) Unix.
3. The method of data transfer – as explained on page 193, the available methods are:
   (a) timed programmed input/output;
   (b) programmed I/O with software polling;
   (c) programmed I/O with interrupt;
   (d) single channel direct memory access (DMA);
   (e) dual channel DMA.

   For even higher acquisition rates such as required in processing video signals, the acquisition board may be connected directly to a digital signal processing (DSP) board instead of relying solely on the main processor in the computer. Examples of this technique are described by Pleau (1993) and Andrews (1993).
4. Processing – having decided how to send data to the processor, we then have to consider what we want (or might want in the future) to do with it. The main options are:

   (a) read to file only – processing is either not required or is written by the user;
   (b) tabulate and plot;
   (c) display graphically or pictorially;

   (d) fit an equation to the measured data;
   (e) Perform statistical tests on the data;
   (f) Perform time-series analysis on the data.
5. Is the analysis required in real time?
6. The format for data input and output:
   (a) ASCII files;
   (b) binary files;
   (c) spreadsheet format;
   (d) database format.
7. Commands – is there a preference for written commands, as in MS DOS, or with icons, as in Windows?

12.3 Driver programs

12.3.1 Drivers for commercial data acquisition boards

All commercial data acquisition boards have **driver** programs available which may be included in the purchase price or may be charged separately.

A driver takes commands from a high-level language such as Pascal, BASIC or C and translates them into the **addresses** and **commands** required to operate a particular board. It also reads the data gathered by the board into variables in the program. Each board has a base memory address set by a DIP switch or jumpers, and the registers on the board which store commands and data occupy a block of memory starting with the base address. It is necessary to allow the user to alter the base address because several identical boards could be plugged into one computer, each taking data from a different source. Each board would be set to a different base address so that the program could distinguish between the different data sources.

In the plug and play system, the boards are configured entirely in software to avoid the fiddly task of setting the DIP switches.

12.3.2 Drivers for GPIB interface boards

A GPIB or IEEE-488 board also requires a driver program. IEEE-488.1 specified all the hardware and software to transfer a single byte of data, but the sequence of bytes to fulfil a particular function was left to the instrument designer. Thus a program which set up and read voltages from one manufacturer's digital voltmeter might give incorrect results or not work at all with another manufacturer's instrument. IEEE-488.2 and the programming language SCPI have standardized the commands as well. GPIB drivers are now designed to use SCPI commands.

12.3.3 Writing drivers

Writing a driver program requires an exact specification of the commands and data of the board being driven, and the interface bus of the particular computer in use. Several variants of the interface bus are described in Chapter 11.

Intel processors use separate instructions for I/O and for addressing memory. A predefined space is available for I/O. To write a driver it will probably be necessary to refer to a technical reference manual for the computer; the user's manual will almost certainly be inadequate. A paper giving many practical details of interfacing and examples of drivers including interrupt routines was published in *Electronics and Wireless World* (anonymous, 1989).

The commands in the high-level language which can operate on specific memory addresses are also needed. Turbo Pascal has the predefined arrays 'Port' and 'PortW' to access the data ports of a processor in the 80 x 86 series and also 'Mem' and 'MemW' which are used to access memory addresses. For example, k: = Port[$300] reads the high byte in address 300 to the variable *k*. In BASIC the instruction PEEK(XXXX) reads the data at address XXXX and POKE(YYYY) writes data to address YYYY.

Writing a driver to run under Windows is more complicated than one for DOS and requires a knowledge of how Windows operates (Barker, 1992; Fountain, 1992). In particular, Windows may not always be ready for data immediately the A/D has finished conversion, so boards intended for use with Windows often include a temporary storage buffer.

12.4 Data presentation

12.4.1 Spreadsheets

Once the incoming data has been acquired, it may be stored in a file or a spreadsheet ready for processing or display. For many purposes, the plotting facilities included in a spreadsheet may be adequate, and are likely to be available at minimum expense.

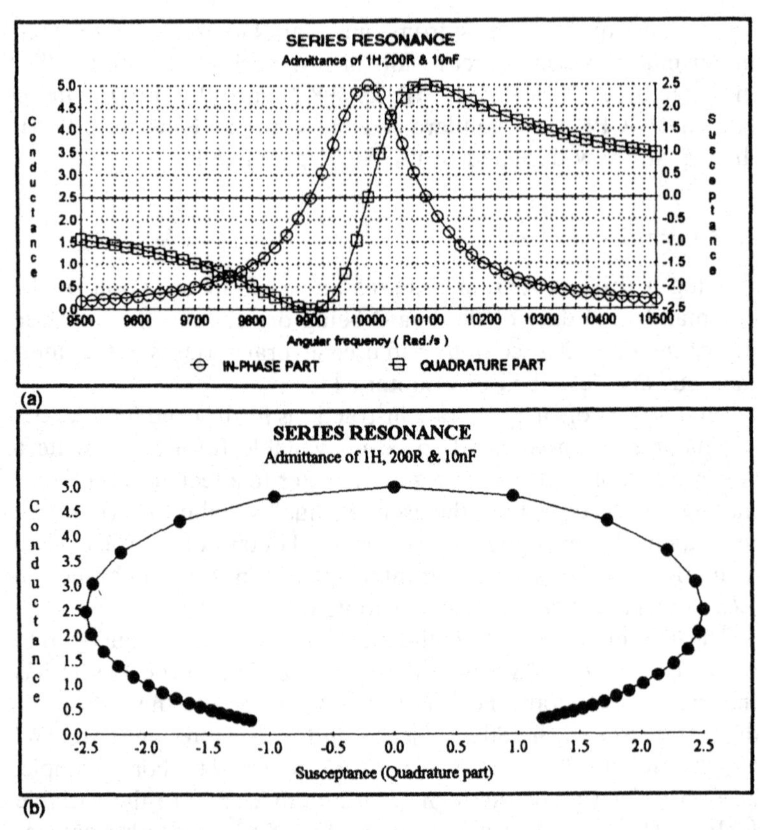

Figure 12.1 *Graphs plotted by a spreadsheet program. (a) In-phase (conductance) and quadrature (susceptance) components of the admittance of a series-resonant circuit plotted against frequency. Inductance = 1 henry; resistance = 200 ohms; capacitance = 10 nanofarads. Note vertical lettering on y-axis and 200R because Ω symbol was not available. (b) The same data as above, but in-phase component (conductance) plotted against quadrature component (susceptance) for a series-resonant circuit. Frequency increases clockwise round the curve.*

Bissel (1994) has shown how useful the computation facilities of a spreadsheet are for engineering applications. Examples of two presentations of the same data plotted by Works for Windows are shown in Figure 12.1.

Spreadsheets will produce Cartesian plots (x–y or x–t) on linear or logarithmic scales with and without joining the points and with limited facilities for labelling the axes. Spreadsheet terminology is derived from business software so a graph is known as a chart and an axis is a category. General acquisition programs and maths programs include more sophisticated plotting facilities.

12.4.2 Graphics packages

There are many graphics packages designed to run on a PC with DOS or Windows, and no doubt there will soon be many more. They can be classified into business packages such as Harvard graphics, scientific packages such as Stanford graphics and painting and drawing packages. Earnshaw and El-Haddadeh (1994) have compared Origin, Stanford graphics, CorelCHART, SigmaPlot and DeltaGraph. Balmforth (1995) has reviewed DeltaGraph, and Adler 1995) has compared Axum and Origin. Earle (1992) has reviewed image processing packages.

Scientific graphics packages give much more flexibility in the presentation of data than spreadsheets. In addition to x–y and x–t graphs, they can present polar plots for functions of angle. Some scientific packages can do Smith charts as well. Also, three-dimensional data can be presented in a variety of ways: contour plots (like a map), surface plots and waterfall plots. A waterfall plot is a sequence of rectangular plots each displaced slightly to the left of and below the previous one. These are particularly useful for data which is a function of two variables. For example, speech may be presented as a parameter of both frequency and time.

The main difference between business and scientific graphics is that scientific packages incorporate some data analysis. The selection of analysis procedures varies from one package to another.

12.5 Data analysis

12.5.1 Introduction

The standard techniques which scientists have used for years for analysing data are now all available in commercial software. A good description of the analysis of experimental data, written for engineers rather than mathematicians, is Holman (1993). A comparative review of Asystant, DADiSP, SigmaPlot, TableCurve and Peakfit was published by Barton (1992) and, no doubt, similar articles will appear from time to time as new programs come on the market. Mathematical software was reviewed by Hereman (1995). A user trying to select a suitable program is well advised to consult the most up-to-date comparison available and/or one of the specialist dealers. Bear in mind that a manufacturer's enthusiastic description of his product will list the facilities available and keep quiet about those which are lacking.

Examples of analytical procedures are given below.

12.5.2 Curve fitting

Curve fitting can be very useful when the acquired data is known to follow a certain equation whose parameters we wish to find.

The simplest equation is that of a straight line, and the 'best' straight line is found as follows. Consider the series of N experimental points and postulate the equation $y = ax + b$. The best line is defined by the values of the slope a and the intercept on the y-axis, b, for which the root mean square of the distances between each point and the line is minimized. When a and b have been found, it is only a small step to compute the value of the rms error which is quoted as a measure of the goodness of fit. Note that this procedure is excellent for removing the effects of random noise, but cannot distinguish between a slight, smooth curve and a straight line with superimposed noise. The method for calculating the 'best straight line' is given in Appendix B.

All the scientific plotting and analyses packages can compute the best straight line (linear regression) or the best polynomial function through a number of recorded data points, and most of them can find the parameters of the best fit for a number of other equations as well (Glock, 1995).

12.5.3 Correlation

The data points acquired from each input channel form a time-series. Successive sections of two series may be compared by the function known as the correlation. The cross-correlation of $x(t)$ and $y(t)$ is illustrated in Figure 12.2.

Each point in the x series is multiplied by each point in the y series delayed by a time τ, and the products are integrated from time = 0 to maximum, T. This gives a value of the correlation function R_{xy} for delay or lag = τ. The process is repeated for all values of τ and the result is the correlation function shown in the third line of Figure 12.2(a).

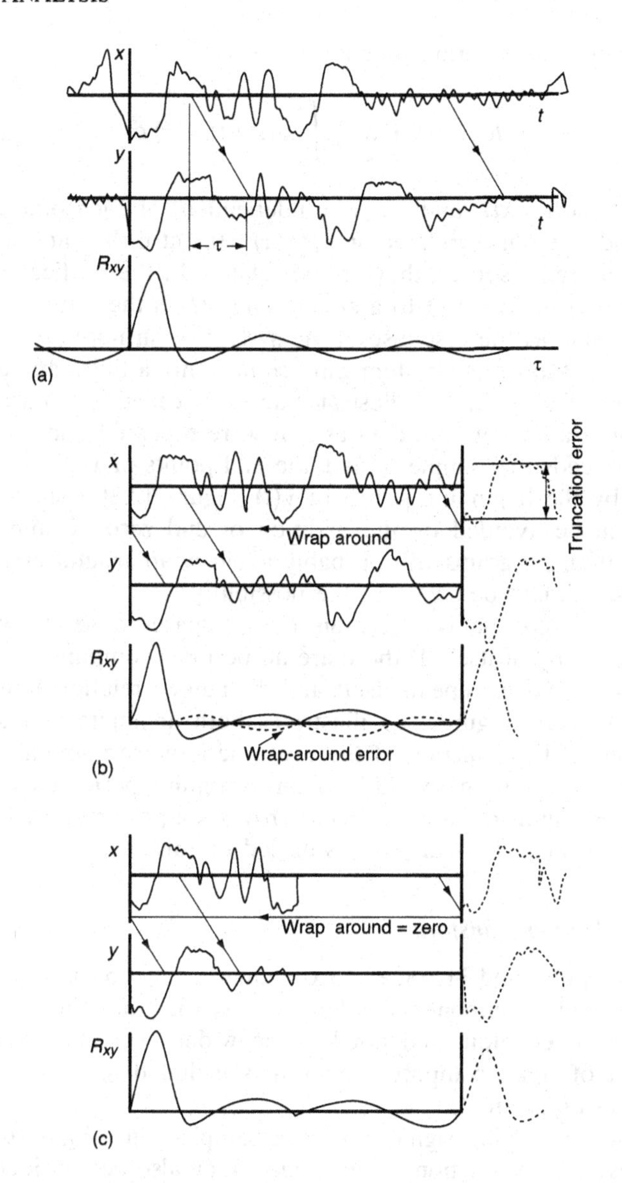

Figure 12.2 *Cross-correlation: (a) continuous functions; (b) finite samples, showing wrap around; (c) finite samples, avoiding wrap-around error.*

The cross-correlation function is

$$R_{xy}(\tau) = \lim_{T\to\infty} \frac{1}{T} \int_0^T x(t)\, y(t+\tau)\, \mathrm{d}t \tag{12.1}$$

If the series $x(t)$ and $y(t)$ are independent of each other, they are said to be uncorrelated and $R_{xy}(\tau) = 0$, but if they are derived from the same source, they are correlated. In the particular case where $x(t)$ is the input to a system and $y(t)$ is the output, $R_{xy}(\tau)$ is the same as the response of the system to an impulse.

A data acquisition system can acquire only a finite amount of data and Figure 12.2(b) illustrates an error called 'wrap around'. The program treats the data as if it were repeated and so when $t + \tau$ exceeds the sample period, the end points of $x(t)$ are multiplied by the beginning points of $y(t)$. Figure 12.2(c) shows how this can be avoided by the addition of end zeros. Commercial correlation programs will probably avoid wrap-around error but the user should be aware of the possibility.

The autocorrelation function R_{xx} compares a series with a delayed copy of itself. If there are no periodic components in the function, it never repeats itself, and the autocorrelation soon dies away to zero. Figure 12.3 illustrates both analogue and digital versions of this function. If there is a periodic component in the original function, there will be a corresponding periodic component in R_{xx}, and so autocorrelation provides a powerful method of detecting weak periodic signals buried in noise.

12.5.4 Fourier transforms

Fourier published his idea of expanding a function into a series of sinusoidal components as long ago as 1822, but this powerful analytical technique did not become widely available until the advent of small computers. Now it is included in almost every analysis program.

When a varying signal has been sampled and digitized, it is expressed as a function of time, but it may also be considered as a function of frequency, and the Fourier transform relates the two. It is particularly useful for computing the frequency response of a system when the input and output have been measured as functions of time.

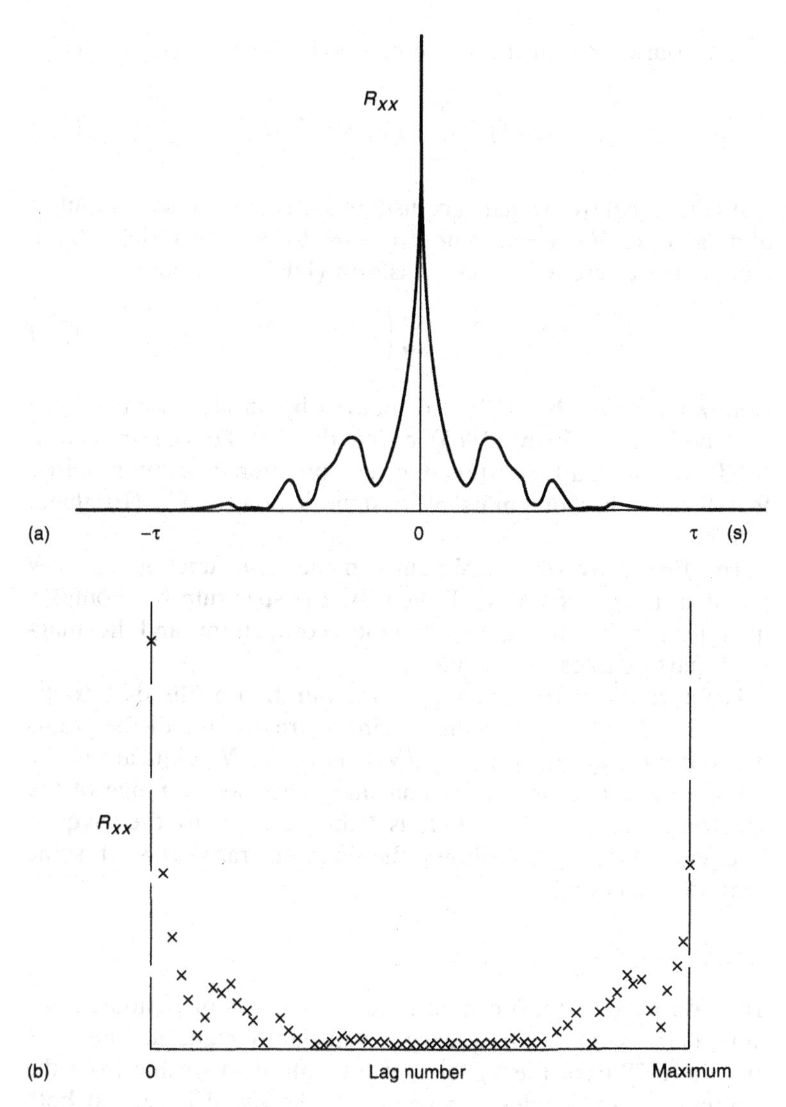

Figure 12.3 *Autocorrelation: (a) typical autocorrelation function for positive and negative lags; (b) discrete representation of (a).*

The Fourier transform of a nonperiodic function is given by

$$x(f) = \int_{-\infty}^{+\infty} x(t)\, \mathrm{e}^{-j2\pi ft}\, \mathrm{d}t \qquad (12.2)$$

As the signal from a data acquisition system comprises a number of data points N, measured not from $-\infty$ to $+\infty$, but during a finite time T, the discrete Fourier transform (DFT), is used:

$$X_k = \frac{1}{N} \times \sum \left(x_n\, \mathrm{e}^{-2\pi nk/N} \right) \qquad (12.3)$$

from $n = 0$ to $n = \mathrm{N} - 1$. It is computed by an algorithm devised by Cooley and Tukey (1965) called the fast Fourier transform (FFT), which greatly shortened the computation time, but required that the number of points N must be a power of 2 (Brigham, 1988).

The FFT converts the N points in the time function x_n, to N points in the spectrum X_k. In general, the spectrum is a complex quantity, in which the real part denotes cosine terms and the imaginary part denotes sine terms.

For a sampling frequency f_{SAM}, the duration of the data transformed is $T = N/f_{\mathrm{SAM}}$ and the spacing, or resolution, of the points in the frequency domain $= f_{\mathrm{SAM}}/N$. Half of the N points are in the real part and half are in the imaginary part, so the range of the spectrum $X(f)$ is $\frac{1}{2} f_{\mathrm{SAM}}$, that is from zero up to the Nyquist frequency. Figure 12.4 shows the Fourier transforms of some common functions.

Windowing

The Fourier transform can be used for samples of periodic functions if the sampling frequency is sufficiently high, but the DFT and the FFT treat the signal as if the data points came from the sampling period endlessly repeated. If the signal is zero at both ends of the sample period, this does not matter. If it is not zero, as indicated by the solid line in Figures 12.4(a) and (b), an error is introduced because the signal is treated as if it jumped suddenly from the last point back to the first point and a sudden jump implies the presence of high frequencies.

This type of error can be avoided by a procedure called windowing. The input signals are multiplied by a function which reduces the signal gradually to zero at each end. Many functions

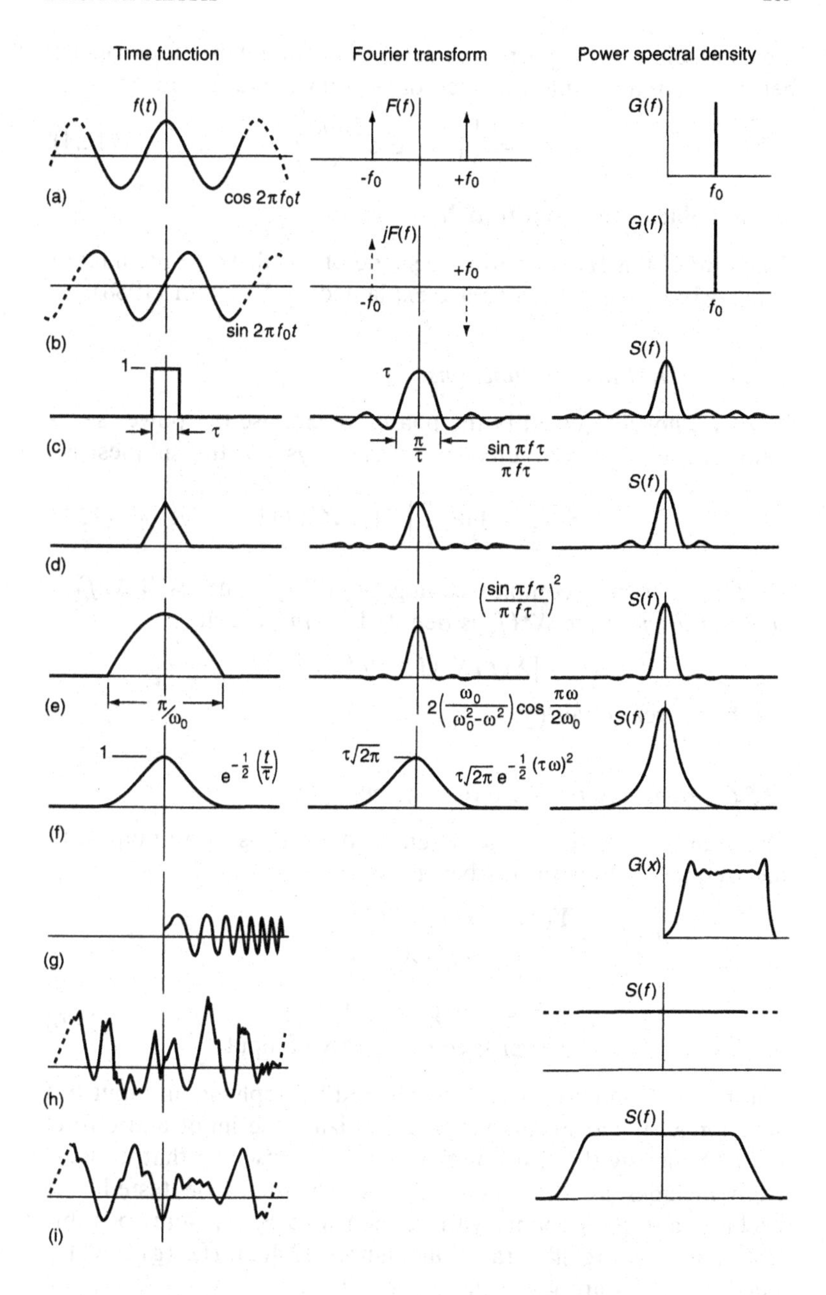

Figure 12.4 *Common Fourier transforms: (a) cosine; (b) sine; (c) square pulse; (d) delta function; (e) half-cosine pulse; (f) Gaussian pulse; (g) fast-swept sine wave; (h) white noise = random; (i) pink noise = band-limited noise.*

are available for this purpose; one of the commonest is the cosine bell or 'Hanning' window. Each data point is multiplied by

$$W_n = \frac{1}{2}\left(1 - \cos\frac{2\pi n}{N}\right) \tag{12.4}$$

for all values of n from 0 to $N-1$.

(The plot of this function has the shape of a bell; hence the name.) Other windowing functions are explained by Brigham (1988).

12.5.5 Spectral density functions

If we are not interested in the phase, we can use the power spectrum. This can be computed in several ways, but the simplest is

$$S(f) = \lim_{T\to\infty} \frac{1}{T}\left[X(f)\,X^*(f)\right] \tag{12.5}$$

$X^*(f)$ denotes the complex conjugate of $X(f)$. That is, if $X(f)$ is denoted $a + jb$, then $X^*(f)$ is denoted $a - jb$, so that

$$[X(f)X^*(f)] = a^2 - b^2$$

which is a real quantity.

12.5.6 Frequency response of a system

The transfer function of a system is defined as output/input. In the frequency domain this becomes

$$H(f) = \frac{Y(f)}{X(f)} = \frac{Y(f)\,X^*(f)}{X(f)\,X^*(f)}$$

$$= \frac{\text{cross-spectral density}}{\text{power spectral density of input}} \tag{12.6}$$

The transfer function may be measured by applying an input test function $x(t)$, and measuring and digitizing the input and output data. Obviously the test function must be chosen so that its spectrum includes all the frequencies which are to be tested, and excludes any components with frequencies $> \frac{1}{2}f_{SAM}$. Some possible test functions are illustrated in Figures 12.4(d), (f), (g) and (i). Further details are given in Taylor (1977).

12.5.7 Statistics

Scientific plotting packages will compute the more common statistical procedures such as the mean, standard deviation and the variance of a set of data and will plot its histogram. Data acquisition packages will do tests for confidence levels and for correlation. Barton (1991) compared the facilities of acquisition packages available at that time.

In addition there are specialized packages for statistics and Currall (1994) has compared nine of the 300 or so on the market. Many of these had been written with medical, biological or sociological applications in mind. Currall concluded that most packages could do the basic procedures listed above together with hypothesis tests, goodness-of-fit tests, ranking tests and linear and multiple regression, but that the packages differed widely in their ability to handle nonlinear relationships between variables. A more general description of the uses of statistics packages was given by Dawn (1995).

12.6 Complete data acquisition packages

12.6.1 Introduction

Packages such as Asyst, Asystant +, DADiSP, LabTech Notebook and TestPoint combine acquisition, analysis and presentation. Prices range from a few hundred to several thousand pounds, depending on the range of facilities offered.

12.6.2 Virtual instruments

A computer equipped with one or more data acquisition boards and a comprehensive analysis and presentation program becomes a virtual instrument. For example, if the PC is acting as a CRO, the screen not only displays waveforms, but icons of the standard CRO controls so that changes in the gain or time-base settings can be made with a mouse. Different software can make the PC behave as a completely different instrument. We can expect rapid developments and wide usage in this area, possibly making today's instruments as obsolete as teak boxes and brass terminals.

Well-established programs of this type include LabVIEW, LabWindows, TestPoint and VEE (virtual engineering environment). A comparative review has been published by Taylor (1995).

LabVIEW and TestPoint are icon-based graphical programming tools, whereas LabWindows is a C or BASIC code generator and therefore needs some programming skill. They all include very large libraries of driver programs for GPIB, VXI, RS-232 and data acquisition boards from various manufacturers. They also include many functions for curve fitting, digital filtering, statistical and numerical analysis. A guide to LabVIEW, in addition to the manufacturer's literature, is Johnson (1994).

VEE is described as a graphical test language, and is icon-based. It was designed for use on workstations, but is now available for PCs running Windows. As you might expect from its origin, it is a very large program with many analytical routines.

With any of these programs, measurement systems can be designed and built from any acquisition boards which can communicate with the host computer by GPIB, RS-232 or VXI buses. As would be expected, these very versatile programs require several megabytes of memory and preferably an 80846 or Pentium processor. They exist in several versions:

- The most comprehensive version is intended for designers of complicated acquisition systems and allows the user to design his experiment, choose all the units in the system, decide parameters like sampling rates, and to process and present the resulting data. The units may exist on plug-in boards in the PC, in a separate chassis linked to the computer by the VXI, GPIB or RS-232 bus, or as independent instruments linked by the GPIB or RS-232 bus.
- Smaller versions include only subsets of the available drivers and analysis routines.
- Run-time versions allow the user to load and run a virtual instrument, but not to modify it.
- Demonstration discs are available free of charge, and demonstrate the range of analysis, control and display facilities using simulated data. Neither a GPIB board or a data acquisition board is required to use a demonstration disc.

The comprehensive version of the package would suit an instrument manufacturer, who could develop virtual instruments. These would comprise the run-time version of the program together with the appropriate plug-in board(s) and their drivers for the end-users to run on their own PCs.

APPENDIX A

Table of ratios and equivalent word lengths

Ratio	*Inverse ratio*	*No. of bits*	*Decibels*
1	1.0000	0	0
2	0.5000	1	6.020 599 91
4	0.2500	2	12.041 199 83
8	0.1250	3	18.061 799 74
10	0.1000		20
16	0.0625	4	24.082 399 65
32	0.03125	5	30.102 999 57
64	0.015625	6	36.123 599 48
100	0.01000		40
128	0.0078125	7	42.144 199 39
256	0.003906	8	48.164 799 31
512	0.001953	9	54.185 399 22
1 000	0.001000		60
1 024	0.0009766	10	60.205 999 13
2 048	0.0004883	11	66.226 599 05
4 096	0.0002441	12	72.247 198 96
8 192	0.0001221	13	78.267 798 87
10 000	1.0000E-04		80
16 384	6.1035E-05	14	84.288 398 79
32 768	3.0518E-05	15	90.308 998 7
65 536	1.5259E-05	16	96.329 598 61
100 000	1.0000E-05		100
131 072	7.6294E-06	17	102.350 198 5
262 144	3.8147E-06	18	108.370 798 4
524 288	1.9074E-06	19	114.391 398 4
1 000 000	1.0000E-06		120
1 048 576	9.5367E-07	20	120.411 998 3

APPENDIX B

Linear regression or finding the best straight line

The best straight line through a number of points is found as follows. Consider the series of n points

$$x_1, y_1; \quad x_2, y_2; \quad x_3, y_3 \ldots x_n, y_n$$

and postulate that the equation of the best straight line is $ax + b$.

The distance of point i from the line is

$$y_i - (ax_i + b)$$

which is the error for this point. We then have to find the values of a and b which give the least value for the root mean square error (minimizing the square of the error instead of the error itself allows for points being above or below the line). The sum of the square of the errors is

$$S = \sum [y_i - (ax_i + b)]^2$$

Differentiating with respect to a and b and equating to zero gives the equations

$$nb + a \sum x_i = \sum y_i$$

$$b \sum x_i + a \sum x_i^2 = \sum x_i y_i$$

Solving these equations gives

$$a = \frac{n \sum x_i y_i - \left(\sum x_i\right)\left(\sum y_i\right)}{n \sum x_i^2 - \left(\sum x_i\right)^2}$$

and

$$b = \frac{\left(\sum y_i\right)\left(\sum x_i^2\right) - \left(\sum x_i y_i\right)\left(\sum x_i\right)}{n\sum x_i^2 - \left(\sum x_i\right)^2}$$

Having found a and b, it is only a small extra step to compute the minimum value of S is $\sqrt{(S/n)}$.

S is the rms error which is quoted as a measure of the 'goodness of fit'.

Note that this procedure is excellent for removing the effects of random noise, but cannot distinguish between a slight, smooth curve and a straight line with superimposed noise.

APPENDIX C

Dielectric comparison chart

Reproduced by courtesy of AVX Corporation, 750 Lexington Ave., New York, NY 10022-1208.

Characteristics		Multi-Layer Ceramics			Ceramic Discs	Internal Barrier Layer	Reduced Titanates	Multi-Layer Glass
		NPO	Stable	HiK				
Capacitance	Range, μF	1pF – .01μF	1pF – 2.2μF	.001 – 10μF	1pF – 0.1μF	.01 – .22μF	.01 – 1.0μF	0.5p – .01μF
	Min Tol. Avail.%	±0.5%	±5%	±20%	Same as Multi-layers	±20%	±20%	±0.5
	Standard Tolerance %	±5%, ±10%	±10%	+80%, −20%	Same as Multi-layers	+80%, −20%	+80%, −20%	±1%, ±5%
Voltage Range	Typical, V_{DC}	50–200	50–200	25–100	50–10,000	50	3–50	50 – 2000
Temperature	Range, °C	−55°C, +125°C	−55°C, +125°C	+10°C, +85°C and −55°C, +85°C	−55°C, +85°C	−55°C, +85°C	−55°C, +85°C	−75°C, +200°C
	Coefficient %ΔC	±0.3%	±15%	+22%, −56% and +22%, −80%	Same as Multi-layers	±30%	±10%, ±30%	±1.65%
Insulation Resistance	<1.0 μF	10^5 MΩ	10^5 MΩ	10^4 MΩ	Same as Multi-layers	10^4 MΩ	10 MΩ	10^5 MΩ
	>1.0μF, MΩ-μF	N.A.	2,500	1,000	N.A.	N.A.	0.1	N.A.
Dissipation Factor	Percent at 1kHz, %	0.1%	2.5%	3.0%	0.1% to 4.0%	5.0%	5% to 10%	0.2%
Dielectric Absorption	Percent Typical, %	0.6%	2.5%	N.A.	Same as Multi-layers	N.A	N.A.	.05%
Frequency Response	Freq. Response 10 = Best, 1 = Poorest	9	8	8	8	3	2	9
	Max. Freq. (MHz) For ΔC = ±10%	100	10	10	Same as Multi-layers	10	1	100
Stability (1000 Hrs)	Typical Life Test, %ΔC	0.1%	10%	20%	Same as Multi-layers	20%	20%	0.5%
Polarity	Single Cap	N.P.	N.P.	N.P.	N.P.	N.P.	N.P.	N.P.

Characteristics		Multi-Layer Glass-K			Mica	Polyester	Poly-carbonate	Poly-propylene	Poly-styrene	Solid Tantalums	Aluminium Electrolytics
		'T' Charac-teristic	'U' Charac-teristic	'V' Charac-teristic							
Capacitance	Range, μF	270pF–.02μF	.012–.039μF	.022–.1μF	1pF–.09μF	.001–10μF	.001–10μF	47pF–0.47μF	100pF–.027μF	.01–1000μF	0.5–10^6μF
	Min Tol. Avail.%	±5%	±10%	±10%	±0.5%	±5%	±1%	±0.5%	±0.5%	±5%	±20%
	Standard Tolerance %	±5%, ±10%	±5%, ±10%	±5%, ±10%	±1%, ±5%	±10%	±10%	±5%	±5%	±20%	+100%, –10%
Voltage Range	Typical, V_{DC}	25–50	25–50	25–50	50–500	100–600	100–600	100–600	30–600	6–125	3–500
Temperature	Range, °C	–75°C, +200°C	–75°C, +200°C	–75°C, +200°C	–55°C, +125°C	–55°C, +125°C	–55°C, +125°C	–55°C, +85°C	–55°C, +70°C	–55°C, +125°C	–40°C, +85°C
	Coefficient %ΔC	+2%, –10%	–2%, –15%	+20%, –45%	–.4% +1.8%	±12%	±2%	±2.5%	±1%	±8%	±10%
Insulation Resistance	<1.0 μF	10^4 MΩ	10^4 MΩ	10^4 MΩ	10^2 MΩ	10^4 MΩ	10^5 MΩ	10^5 MΩ	10^6 MΩ	10^2 MΩ	N.A.
	>1.0 μF, MΩ-μF	N.A.	N.A.	N.A	N.A.	10^3	10^4	N.A.	N.A.	10	100
Dissipation Factor	Percent at 1kHz, %	1.0%	1.5%	3.0%	0.1%	2%	1.0%	0.35%	.1%	8% to 24%	8% (at 120 Hz)
Dielectric Absorption	Percent Typical, %	0.1%	0.1%	1.3%	0.3–0.7%	0.5%	0.35%	.05%	.05%	N.A.	N.A.
Frequency Response	Freq. Response 10 = Best, 1 = Poorest	8	8	8	7	6	6	6	6	5	2
	Max. Freq. (MHz) For ΔC = ±10%	100	75	10	100	N.A.	N.A.	N.A.	N.A.	.002	N.A.
Stability (1000 Hrs)	Typical Life Test, %ΔC	5%	10%	20%	0.1%	10%	5%	3%	2%	10%	10%
Polarity	Single Cap	N.P.	N.P.	N.P.	N.P.	N.P.	N.P.	N.P.	N.P.	Polarized	Polarized

Glossary

Absolute accuracy The absolute accuracy of a D/A or A/D is specified by the worst-case error in output expressed in volts; that is, relative to an absolute, external standard.

Accuracy The closeness of the measured value to the true value.

Acquisition time (Sample and hold). The time taken from the sample command until the output voltage follows the input voltage again to within a specified accuracy.

Active filter or sensor A filter or sensor requiring an external power supply.

A/D abbreviation for analogue to digital converter.

Aliasing An error which results in a set of samples which appear to have come from a signal of a completely different frequency from the real one. Aliasing occurs when the sampling frequency is less than twice the signal frequency or any component of it.

Analogue A quantity which is continuously varying, as distinct from a digital quantity which varies in distinct steps. Two quantities which vary in proportion are analogues of each other. Commonly, one is the parameter being measured and the other is electrical.

Analogue to digital converter The device which converts the signal from analogue to digital form.

Anti-aliasing filter A low-pass filter to remove components of the signal liable to cause aliasing errors.

Aperture jitter The variation in the aperture time. This is important where several channels should be held simultaneously. Also called aperture uncertainty.

Aperture time In a sample and hold circuit, the aperture time lasts from the hold command until the switch is fully open. In some designs there may be a further delay until the output has settled.

Aperture uncertainty error The uncertainty in the output of an A/D converter caused by finite aperture time.

ASCII American Standard Code for Information Interchange – a code representing all upper and lower case letters, numerals and punctuation. Originally seven bits, is now extended to eight. Widely used for transmission of alphanumeric information.

Bandgap reference An integrated circuit voltage reference device utilizing the base–emitter voltage characteristics of two transistors.

Bandwidth For a flash converter, bandwidth is defined as the maximum frequency, full-range input sine wave which can be accurately digitized. For a filter or amplifier, it is the range of frequencies within which the gain is within 3 dB of the pass-band gain.

Baud rate The number of bits transferred per second. Some of these bits may be control bits, in which case the baud rate will be greater than the data transfer rate; e.g. in RS-232 each word comprises one start bit, eight data bits and two stop bits.

BCD In binary coded decimal each decimal digit is separately coded into a 4-bit binary word.

Bessel filter A filter which has a phase shift in the pass-band which is proportional to frequency, so that all components of the signal are delayed by the same time. This minimizes distortion of a complex waveform.

BIPM Bureau International des Poids et Measures – central office and laboratory at Sèvres near Paris responsible for development, dissemination and maintenance of standards.

Bit A binary digit. The smallest unit of a digital signal. It can have only two states which may represent high or low, off or on, yes or no, 0 or 1.

BSI British Standards Institution, London – UK publisher of standards documents.

Bus A number of wires carrying related signals and each connected to several devices. There are many standard bus systems, some linking a microprocessor to its peripherals, some linking boards and some linking complete instruments.

Butterworth filter This type of filter has the least attenuation in the pass-band, hence its alternative name of maximally flat.

Byte Eight bits which can represent 256 different values.

Calibration Assessing the accuracy of a sensor, measuring instrument or A/D converter by applying a known value of the input and observing the output.

Capture range The range of frequencies to which the phase-locked loop will lock after running free, which is less than the lock range.

CGPM General Conference of Weights and Measures – this organization has responsibility for SI at political level, and meets every few years.

Charge amplifier A type of operational amplifier circuit used with piezoelectric sensors. The output voltage is proportional to the charge, not the voltage, applied at the input.

Charting Computer jargon for plotting a graph.

CIPM Comité International des Poids et Measures – committee of scientists carrying out decisions of CGPM and supervising BIPM.

Clock The source of all the timing pulses needed to operate the units in the correct sequence.

CMRR Common mode rejection ratio – a voltage between a source and the common line should have no effect on a differential amplifier or a DVM, but in practice a fraction of it may appear across the amplifier input terminals. The CMRR is the reciprocal of this fraction and is usually expressed in decibels.

Confidence levels The probability that the correct value lies within specified limits.

Correlation Cross-correlation is a mathematical procedure for comparing two functions of time to find whether they derive from a common source, and to measure any time delay between them. Autocorrelation compares one function with itself and measures the frequency of any periodic components in it.

Critical damping The damping which is just, and only just, sufficient to prevent oscillation when the system receives a transient disturbance.

Crosstalk The leakage of signals between channels of a multiplexer. It is measured in terms of crosstalk attenuation, expressed in decibels.

D/A Abbreviation for digital to analogue converter.

Damping factor Damping force divided by mass.

Damping force In a damped spring-mass system, the damping force is proportional to velocity in the direction opposing motion. Damping force implies force per unit velocity.

Damping ratio The ratio of actual damping force to critical damping force.

Data A series of numbers representing a quantity being measured, transmitted, analysed or recorded. Grammatically, data is the plural of datum, although in technical usage the meaning is slightly different.

Data acquisition The branch of engineering dealing with collecting information from a number of analogue sources and converting it to digital form suitable for transmission to a computer, printer or alphanumeric display.

Data logger An instrument for recording data, usually from a large number of channels over a long period.

Datum The reference value from which all other values are measured.

Dielectric absorption A property of a dielectric such that if a capacitor is charged and then discharged, the charge recovered is less than the charge supplied. Some of the charge appears to have been absorbed by the dielectric.

Differential linearity error In a D/A converter, this is the maximum difference between an actual step and the nominal step for one LSB change in output.

Digital multimeter A digital voltmeter with the addition of circuits for measuring alternating voltage, direct and alternating current and resistance. Many DMMs also include some computation, memory and a bus interface to transmit the measurement to a computer or recorder.

Digital to analogue converter A device for converting a signal from digital to analogue form.

Digital voltmeter An instrument for measuring voltage by digital techniques. It comprises attenuators and amplifiers for various ranges, an A/D converter and a display. The term is sometimes used for instruments which would more correctly be called digital multimeters.

Digitization The process of converting an analogue signal to digital form.

Discrimination Alternative for resolution.

DMA Direct memory access – a high-speed method of transferring data from the data bus to the memory without passing through the processor. Special hardware is needed.

DMM Abbreviation for digital multimeter.

Driver A subroutine which takes commands from a high-level language and translates them into the addresses and commands required to operate a particular board. It also reads the data gathered by the board into variables in the main program.

Droop rate While a sample and hold circuit is in the hold mode, the output may drift slightly. The rate of drift is called the droop rate.

DVM Abbreviation for digital voltmeter.

EIA Electronic Industries Association, USA.

Error The difference between the measured value and the true value of the measurand.

Feedthrough In a sample and hold circuit in hold mode, feedthrough is any fraction of the input voltage which appears at the output. In a multiplexer, feedthrough is any fraction of the input to an off channel which appears at the output.

FET Field effect transistor – may be used as a high-input impedance amplifier or a switch. FETs are subdivided into junction (JFETs), metal oxide semiconductor (MOSFETs) and complementary MOSFETs (CMOS).

FFT Fast Fourier transform – an algorithm for computing the Fourier transform for sampled data. The number of points must be a power of 2.

Flying capacitor A capacitor with neither terminal connected to a fixed potential. It is used for the temporary storage and transfer of a voltage.

Fourier transform A mathematical method of converting a signal from a function of time to a function of frequency or vice versa.

Full-range voltage In a bipolar A/D or D/A converter full-range voltage is the difference between minimum and maximum voltages corresponding to all zeros and all ones code, respectively.

Full-scale voltage In a unipolar D/A or A/D converter, full-scale voltage is the voltage corresponding to all ones code (after allowing for the offset error, if any).

Gain error An error in a D/A or A/D converter which is proportional to the input. Also applies to a sample and hold in sample mode.

Gauge factor The fractional change in the resistance of a strain gauge is greater than the strain it is measuring and the gauge factor is defined as the fractional increase in resistance divided by the strain.

Gaussian distribution Alternative name for normal distribution.

Glitch A transient or spike at the output. In a D/A converter a glitch is caused by the difference in the turn-off and turn-on times of different bits. In a sample and hold it is caused by capacitive coupling to a step change of voltage somewhere else in the circuit.

GPIB General purpose interface bus – a bus system for connecting instruments to each other and to a computer. Specified by IEEE standards 488-1 and 488-2.

Gray code A binary code arranged so that as the value increases there are no transitions where two or more bits change simultaneously.

Handshake A sequence of signals controlling the flow of data between a device and a bus or vice versa. In the GPIB bus they are 'not ready for data', 'data available' and 'not data accepted'. The 'not' in the name only indicates the polarity of the signal.

Histogram A plot with the value of a parameter on the x-axis and the number of occurrences of that value on the y-axis.

HPIB Hewlett-Packard interface bus – the IEEE-488 standard bus on Hewlett-Packard instruments.

IEC International Electrotechnical Commission.

Independent linearity error The maximum deviation between the actual input/output characteristic and the best straight line through it.

Integrating converter An A/D converter which measures the mean voltage over a specified interval. Used to reduce errors due to mains frequency interference.

Interface A circuit to connect one element to another, in particular to connect an A/D or a D/A to a microprocessor bus.

ISO International Standards Organisation.

ITS-90 International Temperature Scale adopted in 1990, and replacing the International Practical Temperature Scale of 1968.

Kelvin The SI unit of temperature, defined as 'the temperature of the triple point of water divided by 273.16'. The Kelvin is numerically the same as the degree Celsius, but is measured from absolute zero, not from the freezing point of water which is 273.15 K.

Kelvin effect Same as Thomson effect (William Thomson became Lord Kelvin).

Linearity error If the output of a device should be proportional to the input, any departure from proportionality causes linearity error. This may occur in a sensor, amplifier, D/A or A/D converter. See also independent and terminal-based linearity errors.

Lock range The range of frequencies which a phase-locked loop will follow once it has locked.

LSB Least significant bit – a change of this bit represents the smallest possible change in the value of the whole word.

Measurand The quantity being measured.

Mesh analysis A method of analysing complex electrical circuits based on the fact that the sum of all the voltage sources and voltage drops around a closed loop is zero (Kirchhoff's second law).

Metrology The science of measurement.

Microprocessor compatible An A/D converter which can be directly connected to a microprocessor without an interface.

Modem An interface to enable data to be transmitted over a pair of dedicated wires, or over the public telephone lines. The name is an abbreviation of modulate–demodulate.

Modulus of elasticity The modulus is defined as 'stress divided by strain', and has the same units as stress. The modulus for longitudinal strain is Young's (symbol Y or E).

Monotonic A D/A converter is monotonic if the output always increases with every increase in input code.

MSB Most significant bit – a change of this bit represents a bigger change in the value of the word than a change of any other bit.

Multiplexer A selector switch connecting one channel at a time to the analogue to digital converter.

MUX Abbreviation for multiplexer.

Natural frequency If a lightly damped spring-mass system receives a transient disturbance, it will vibrate at its natural frequency. Damping reduces the natural frequency slightly.

Nibble Half a byte, i.e. four bits.

NIST National Institute of Standards and Technology – the standards laboratory for the USA.

Nodal analysis A method of analysing complex electrical circuits based on the fact that the sum of all the currents flowing in to and out of a point (node) is zero (Kirchhoff's first law).

Nonintegrating converter An A/D converter which measures the voltage in a short interval as an approximation to an instantaneous measurement.

Nonlinearity See linearity error.

Normal distribution Large errors are less likely than small errors. The 'distribution' is the equation relating the size of the error to the probability of its occurring. Random errors are distributed according to the bell-shaped curve shown in Figure 2.1.

Normal mode US usage for 'series mode'.

NPL National Physical Laboratory – the standards laboratory for the UK.

Nyquist frequency Half the sampling frequency; also known as the cut-off frequency.

Offset The actual output at zero input (or whatever input should give zero output). Applies to sensors, amplifiers and D/A converters.

Offset binary A code in which all zeros represents the maximum negative number and all ones represents the maximum positive number.

Open collector gate Open collector gates have an external resistor in the collector circuit. If all the gates on the bus share the resistor, the bus goes low if the collector of any gate draws current. In positive logic this is 'wired AND' and in negative logic it is 'wired OR'.

Passive filter or sensor A filter or sensor without an external power supply, all the output power being drawn from the input.

Pedestal error The difference between the offset in sample mode and the offset in hold mode.

Peltier effect When current flows from one material to another, heat is absorbed or released at the junction according to the direction of current flow.

Phase-locked loop The output of a phase-sensitive detector (PSD) drives the frequency control of a voltage-controlled oscillator (VCO). This provides the reference to the PSD either directly or after a frequency divider. Used as a variable frequency band-pass filter or as a frequency multiplier.

Phase-sensitive detector A phase-sensitive detector (PSD) or amplifier has both signal and reference inputs. The output is a function of the amplitude of the signal input and the phase angle between the two inputs.

Piezoelectric charge constant The charge generated per unit force applied. The units are $C\,N^{-1} = mV^{-1}$. The charge constant applies to a sensor driving a low input impedance amplifier (charge amplifier). The symbol is d with two subscripts. The first one, i, indicates the electrical direction and the second, j, the mechanical direction.

Piezoelectric voltage constant The electric field (voltage gradient) generated per unit stress applied. The units are $V\,m\,N^{-1} = m\,C^{-1}$. The voltage constant applies to a sensor driving a high input impedance amplifier. The symbol is g with two subscripts which indicate the electrical direction and the mechanical direction, as for the charge constant.

Poisson's ratio When a material is stretched, it not only becomes longer but also narrower. Poisson's ratio quantifies this effect and is defined as 'the fractional decrease in width divided by the fractional increase in length'.

Precision The closeness together of the measured values when the measurement is repeated.

PRT Platinum resistance thermometer.

Quad Part of a D/A converter controlling four bits.

Quantization The process of approximating a continuous (analogue) signal to a finite number of values which can be represented by the digital system in use.

Quantization error The difference between the actual value of the signal and the nearest value which can be represented digitally.

Quantization noise When an analogue signal is digitized, quantization error is introduced. If the digital signal is reconverted to analogue form, the errors appear as noise superimposed on the original signal.

RAM Random access memory.

Random access memory Memory used for the temporary storage of data either in a computer or separately. Each digital word can be written to or read from its own address in any order.

Random error An error whose value is determined by chance. May be reduced by taking the average of many observations; see also standard deviation.

Real time A signal or event is analysed in real time if the analysis is fast enough to keep up with the incoming data indefinitely. For example, if 1000 data points are acquired per second then, in real time, 1000 results are produced per second, albeit possibly delayed.

Reed relay A reed switch comprises two thin, flat strips of ferrous metal sealed into a glass tube which is filled with inert gas. A reed relay is a reed switch operated by the current in an external coil.

Relative accuracy The relative accuracy of an A/D or D/A is specified by the worst-case error expressed as a proportion or percentage of the full scale.

Repeatability The closeness of the measured values when repeated with the same method, observer, instrument, location and conditions. Repeatability is a measure of random error ignoring systematic error.

Reproducibility The closeness of the measured values when repeated with different methods, observers, instruments, locations, conditions and time. Reproducibility takes account of both random and systematic errors.

Resolution The smallest change in the measurand which can be detected. In an analogue system, the resolution is limited by the noise level, but in a digital system it is limited by the size of the least significant bit or the noise level, whichever is larger.

Root mean square (rms) A method of specifying the magnitude of an alternating quantity. Let y be a function of t with period T, then

$$y_{\text{rms}} = \left(\frac{1}{T}\int_0^T y^2 \, dt\right)^{1/2}$$

If the function is sinusoidal, then the rms value is 0.7071 × peak value ($0.7071 = 2^{-1/2}$).

RTD Resistive temperature detector, US usage for platinum resistance thermometer.

Sample and hold circuit In sample mode the circuit is a unity gain amplifier; when the mode changes to hold the output remains constant. Used before the MUX to sample several channels simultaneously or to prevent the input to the A/D changing during digitization.

Sampling converter An A/D converter with a built-in sample and hold.

Sampling rate The number of samples taken in a second from one channel of a multichannel system – compare with throughput. The speed of a flash converter is specified in terms of sampling rate instead of conversion time.

SCPI Standard Commands for Programmable Instrumentation – used with the GPIB bus.

Seebeck effect Whenever there is a temperature gradient along a conductor, there is also a potential gradient. Hence a circuit made of two (or more) different materials generates an emf if the junctions are at different temperatures.

Sensitivity The relationship of the change in response to the change in stimulus.

Sensor A device which converts the quantity being measured (measurand) to a proportional electrical signal.

Series mode A series mode voltage is an extra, unwanted voltage in series with the one to be measured.

Series mode rejection ratio Maximum noise in the input divided by maximum resulting error in the output. Expressed in decibels. Applies to integrating A/D converters.

Settling time Applies to multiplexers and D/A converters. The time for the output to reach and stay within a specified fraction of its final value. For a D/A the specified limit is usually ± ½ LSB for a full-range change of input, but settling time may also be defined for the switching point of the MSB.

SI Système Internationale d'Unités – developed from the metric system, accepted by international agreement and now serves the scientific and engineering communities of all nations.

Sign magnitude A code in which the polarity of the value is represented by the MSB and magnitude (modulus) of the value is represented by the remaining bits.

Signal conditioning The process and equipment for converting the signal from the sensor to the correct level and frequency for the analogue to digital converter. It includes amplifiers, filters and a.c. to d.c. converters.

Slew rate Rate of change of voltage – with reference to an op-amp, the maximum rate of change in output in response to a step function input is implied.

Software polling A method of identifying the source of an interrupt by interrogating each possible source in turn.

Standard (documentary) A document specifying an agreed method or dimensions or both. Standards are particularly useful where the products of several manufacturers are to be used together. Examples are standard bus systems.

Standard (of measurement) The practical representation of a unit – it may be 'material' (as the standard kilogram, which is a particular platinum–iridinum mass), or 'recipe' (as the second, which is defined from the radiation emitted by caesium under specified conditions).

Standard deviation A measure of the repeatability of a measurement, symbol σ, and defined (2.1) as

$$\sigma = \left[\sum_{i=1}^{i=N} \frac{(x_i - x_{\mathrm{m}})^2}{N - 1} \right]^{1/2}$$

where N is the number of readings, x_1 to x_N, and x_{m} is the mean value of the readings. Standard deviation is effectively the root mean square of the deviation of the measured values from the mean value.

Step error Alternative name for 'differential linearity error'.

Strain Strain is a ratio defined as 'extension divided by original length'. In longitudinal strain these are in the same direction. In shear strain they are at right angles. Torsional strain is the angle of twist per unit length of shaft, cylinder etc.

Stress Stress is defined as 'force divided by area.' In longitudinal stress the area is measured normal to the force. In shear stress the force is in the same plane as the area. The unit of stress is the pascal or newton per square metre.

Systematic error An error whose value remains constant when the measurement is repeated under the same experimental conditions, and remains constant or varies according to a definite law where the conditions change (BS 5233). Systematic errors cannot be reduced by averaging.

Temperature coefficient The change in a parameter when the temperature increases by one degree Celsius, expressed as a proportion of the parameter at a reference temperature.

Terminal-based linearity error The maximum deviation between the actual input/output characteristic and the line joining the end points. Also known as **integral linearity error**.

Thomson effect Heat is absorbed or released by a current flowing through a conductor made of uniform material if there is a temperature gradient along the conductor.

Three-state logic Devices used to connect data to a bus have three states: high-level, low-level and high impedance. One device is selected and sets the bus to high or low level. The others are in the high impedance state and have no effect on the bus.

Throughput The total number of samples converted per second in a multichannel system.

Time constant The time taken for the output to reach 63.2% of its final value when the input is applied suddenly, i.e. a step function input.

Track and hold circuit Alternative name for sample and hold circuit.

Two's complement binary A code in which all zeros represents zero analogue value, and all ones represents one LSB less. A negative number is indicated by MSB = 1.

UART Universal asynchronous receiver transmitter – an integrated circuit for converting serial data to parallel and vice versa.

Uncertainty An estimate of the range of values within which the true value of the measurement lies (BS 5233).

Variance The square of the standard deviation.

VIA Versatile interface adaptor. An integrated circuit used to connect A/D or D/A converters to a microprocessor bus. It also includes counting and timing circuits.

Virtual instrument A data acquisition system and a computer programmed to make measurements and display them similarly to a conventional instrument, oscillograph, voltmeter or spectrum analyser, etc. A change in the software replicates an entirely different instrument.

Window function A time function cannot be sampled indefinitely, and if the sampled function begins or ends suddenly, errors are introduced in the Fourier transform. The window function makes the time function start and stop gradually before transformation.

Word Several bytes used together. Thus a 16-bit word comprises two bytes and can represent $256 \times 256 = 65\,536$ different values.

Zener diode A voltage reference diode – if the voltage is increased in the reverse-biased direction, the current increases suddenly at a voltage characteristic of the device.

References

Adler, E. (1995) Technical graphics and data analysis for Windows PCs. *Scientific Computing World*, (9) 25–28.

Analog Devices (1989–90) Sigma–delta converters, *DSPatch*, Nos 14–17, Analog Devices Inc., Norwood, MA.

Andrews, W. (1993) DT-Connect II quintuples earlier transfer rate. *1993 Applications Handbook* Vol 2, No. 1, Data Translation Inc., Marlboro, MA, pp. 145–147.

Anonymous (1989) Data acquisition using the IBM PC. *Electronics and Wireless World*, **95**(1637), 266–270.

Arbel, A.F. (1980) *Analog Signal Processing and Instrumentation*, Cambridge University Press, UK.

Atherton, A. and Fitton, D. (1989) Temperature definition and measurement using platinum resistance thermometers. *Transactions of the Institute of Measurement and Control* **11**(1), 15–24.

Bailey, A.E. (1982) Units and standards of measurement, *Journal of Physics E: Scientific Instruments* **15**(9), 849–856.

Balmforth, C. (1995) DeltaGraph: technical graphing for both PCs and Macs. *Scientific Computing World*, (13), 33–34.

Bannister, B.R. and Whitehead, D.G. (1991) *Transducers and Interfacing*, Chapman and Hall, London.

Barker, J. (1992) Writing a device interface for Windows. *BYTE*, **17**(7), 303–306.

Barney, G.C. (1988) *Intelligent Instrumentation*, 2nd edn, Prentice-Hall, London.

Barton, D. (1991) A PC user's guide to data acquisition. *Physics World*, **4**(2), 56–60.

Barton, D. (1992) A P.C. user's guide to data analysis. *Physics World*, **5**(2), 41–44.

Bissel, C.C. (1994) Spreadsheets in the teaching of information engineering. *Engineering Science and Education Journal*, **3**(2), 89–96.

Bolger, S. (1980a) Rely on IC-analog switches for fast small-signal control, *Electronic Design News*, 5 August, 105–111.

Bolger, S. (1980b) IC-analog-switch limitations need not hamper designs, *Electronic Design News*, 20 August, 122–130.

Borrill, P. (1987) IEEE 896.1: the Futurebus. *Electronics and Power*, **33**(10), 628–631.

Bowron, P. and Stephenson, F.W. (1979) *Active Filters for Communication and Instrumentation*, McGraw-Hill, New York.

Brigham, E.O. (1988) *The Fast Fourier Transform and its Applications*, Prentice-Hall, Englewood Cliffs, NJ.

Brignell, J. and White, N. (1994) *Intelligent Sensor Systems*, IOP Publishing, Bristol, UK.

British Standards Institution (1973 confirmed 1981) *International thermocouple reference tables*. BS 4937: Parts 1–4, BSI, London.

British Standards Institution (1974 confirmed 1981) *International thermocouple reference tables*. BS 4937: Parts 5–7, Equivalent to IEC 584.1, BSI, London.

British Standards Institution (1984) *Specification for industrial platinum resistance thermometers*. BS 1904:1984, BSI, London.

British Standards Institution (1986 confirmed 1993) *Glossary of terms used in metrology*. BS 5233:1986, BSI, London.

British Standards Institution (1995) *Vocabulary of metrology*. PD6461:Part 1, BSI, London.

Brokaw, A.P. (1984) *An IC amplifier user's guide to decoupling, grounding and making things go right for a change*, Application Note 202, Analog Devices, Norwood, MA.

Caristi, A.J. (1989) *IEEE-488; General Purpose Instrumentation Bus Manual*, Academic Press, London.

Carter, P.W. (1995) Equipment calibration and quality control. *Engineering Science and Education*, **4**(5), 207–210.

Chenhall, H. (1987) Seeking the ultimate in A/D precision. *Electronic Product Design*, December, 39–42.

Clayton, G.B. (1982) *Data Converters*, Macmillan Press, London.

Connelly, J.A. (1973) A general analysis of the phase locked loop, *Electronic Engineering News*, July/August, 37–41.

Cooley, J.W. and Tukey, J.W. (1965) An algorithm for the machine calculation of complex Fourier series. *Mathematics of Computing*, **19**(90), 297–391.

Cronin, L.B. (1995a) Calibration awareness. *Engineering Science and Education*, **4**(5), 194–195.

Cronin, L.B. (1995b) The evolution of quality control standards for calibration systems over the past thirty years. *Engineering Science and Education*, **4**(5), 196–200.

Currall, J. (1994) Statistics software: a means to an end. *Physics World*, **7**(10), 44–48.

Cutkosky, R.D. (1974) New NBS measurements of the absolute farad and the ohm. *IEEE Transactions on Instrumentation and Measurement*, **23**(4), 305.

Daniel, V.V. (1967) *Dielectric Relaxation*, Academic Press, London.

Datron (1987) *Datron Instruments 1281 Seminar*, Datron Instruments, Norwich, UK.

Dawn, T. (1995) Computers transform scientists' use of statistics. *Scientific Computing World*, (8), 21–26.

Dettmer, R. (1987) STE goes up the wall. *Electronics and Power*, **33**(10), 637–639.

Dijkmans, E.C. and Naus, P.J.A. (1987) An experimental 16-bit A/D converter for digital audio applications, IEE Digest No. 1987/48, p. 5/1.

Doebelin, E.O. (1990) *Measurement Systems: Applications and Design*, 4th edn, McGraw-Hill, New York.

Earle, G. (1992) A user's guide to image processing. *Physics World*, **5**(8), 31–36.

Earnshaw, R. and El-Haddadeh, B. (1994) PC graphics: what to look for. *Physics World*, **7**(5), 47–51.

Edwards, R. (1987) Microcomputer buses. *Electronics and Power*, **33**(10), 624–627.

Fairhead, H. (1992) *The 386/486 PC; A Power User's Guide*, 2nd edn, I/O Press, Leyburn, UK.

Fountain, T. (1992) Engineering with 'Windows'. *IEE Review*, **38**(11), 377–379.

Frank, R. (1996) *Understanding Smart Sensors*, Artech House, London.

Gann, R. (1994) VL-BUS v PCI: the local bus debate. *What Personal Computer?*, (56), 70–75.

Gardner, F.M. (1979) *Phaselock Techniques*, 2nd edn, Wiley, New York.

Gardner, J.W. (1994) *Microsensors*, Wiley, Chichester.

Gardner, J.W. (1995) Microsensations. *IEE Review*, **41**(5), 185–188.

Givens, S. (1979) Reduce crosstalk errors in analog multiplexers. *Electronic Design News*, 5 October, 103–109.

Glock, B. (1995) Software chooses equations to fit data. *Scientific Computing World*, (5), 33–38.

Glover, M., Hermes, B., Kup, B. *et al.* (1991) *A fully integrated bitstream analogue-to-digital converter for digital audio*, International Conference on Analogue to Digital and Digital to Analogue Conversion, Swansea, IEE Conference Publication No. 343, pp. 165–167.

Goeke, W.C. (1989) An 8½ digit integrating analog-to-digital converter with 16-bit, 100,000-sample-per-second performance. *Hewlett-Packard Journal*, **40**(2), 8–15.

Göpel, W., Hesse, J. and Hemel, J.N. (1989) *Sensors: A Comprehensive Survey* (six volumes), VCH Publishers, Cambridge, UK.

Gordon, B.M. (1978) Linear electronic analog/digital conversion architectures, their origins, parameters, limitations, and applications. *IEE Transactions on Circuits and Systems*, **25**(7), 391–418.

Hartland, A. (1988) Quantum standards for electrical units. *Contemporary Physics*, **29**(5), 477–498.

Hereman, W. (1995) Visual data analysis; maths made easy. *Physics World*, **8**(4), 49–53.

Hewlett-Packard (undated) *Dynamic performance testing of A-to-D converters*, Product Note 5180A-2, Hewlett-Packard Company, Palo Alto, CA.

Hewlett-Packard (1989) *Floating measurements and guarding*, Application note number 123, 2nd edn, Hewlett-Packard Company, Palo Alto, CA.

Holman, J.P. (1993) *Experimental Methods for Engineers*, 6th edn, McGraw-Hill, New York.

Horowitz, P. and Hill, W. (1989) *The Art of Electronics*, 2nd edn, Cambridge University Press, UK.

Hoskins, K. (1985) Clock-controlled anti-aliasing filter. *Electronic Product Design*, **6**(11), 47–53.

IEEE (1987) *IEEE Standard 488.1-1987 IEEE Standard Digital Interface for Programmable Instrumentation (Reaffirmed 1994)*, Institute of Electrical and Electronics Engineers, Piscataway, NJ.

IEEE (1990) *Standard Commands for Programmable Instrumentation*, Institute of Electrical and Electronics Engineers, Piscataway, NJ.

IEEE (1992) *IEEE Standard 488.2-1992 Standard Codes, Formats, Protocols and Common Commands*, Institute of Electrical and Electronics Engineers, Piscataway, NJ.

IOtech (1991) *Instrument Communication Handbook*, IOtech, Cleveland, OH.

Jaffe, H. and Berlincourt, D.A. (1965) Piezoelectric transducer materials. *Proceedings of IEEE*, **53**(10), 1372–1385.

Johnson, G.W. (1994) *LabVIEW Graphical Programming Techniques*, McGraw-Hill, New York.

Johnston, J. (1991) *New design techiques yield low-power, high-resolution delta–sigma and SAR ADCs for process control, medical, seismic and battery powered applications*, International Conference on Analogue to Digital and Digital to Analogue Conversion, Swansea, IEE Conference Publication No. 343, pp. 118–123.

Jones, O.C. (Chairman) (1989) *Colloquium on changes in the values of the UK national reference standards for the volt and the ohm*, IEE Digest No. 1989/7.

Jones, R.G. (1989) *Changes in the values of the UK reference standards of electromotive force and resistance*, National Physical Laboratory, Teddington, UK.

Kaye, G.W.C. (1986) *Tables of Physical and Chemical Constants*, 15th edn, Longman, London.

Keithley Test Instrumentation Group (1995) *Switching Handbook*, 3rd edn, Keithley Instruments, Cleveland, OH.

Kibble, B.P. (1986) The SI ampere and the volt. *International Journal of Electrical Engineering Education*, **23**(4), 293–302.

Kibble, B.P., Robinson, I.A. and Bellis, J.H. (1988) *A realisation of the SI watt by the NPL moving-coil balance*, Report DES 88, National Physical Laboratory, UK.

Knoblock, D.E., Loughry, D.C. and Vissers, C.A. (1975) Insight into interfacing, *IEEE Spectrum*, **12**(5), 50.

Kuijk, K.E. (1973) A precision reference voltage source. *IEEE Journal of Solid-state Circuits*, **8**(3), 222–226.

Locke, D. (1987) *Digitising in the gigahertz range*, Colloquium on advanced A/D conversion techniques, IEE Digest No. 1987/48, pp. 10/1–10/4.

Mar, A. and Regimbal, D. (1990) Mixed-signal processor merges DSP with A/D, D/A converters, *Analog Dialog*, **24**(2), 3–7.

Mazda, F.F. (1987) *Electronic Instruments and Measurement Techniques*, Cambridge University Press, UK.

McLeod, D.A. (1991) *Dynamic testing of analogue to digital converters*, International Conference on Analogue to Digital and Digital to Analogue Conversion, Swansea, IEE Conference Publication No. 343, p. 29.

Meade, M.L. (1982) Advances in lock-in amplifiers, *Journal of Physics E: Scientific Instruments*, **15**(4), 395–403.

Middlehoek, S. and Audet, S.A. (1989) *Silicon Sensors*, Academic Press, London.

Moralee, D. (1995) PCI: Scientific computing's bus of the future. *Scientific Computing World*, (12), 11–14.

National Physical Laboratory (1989) *Units of Measurement*, 5th edn, NPL Poster Dd 8171158, National Physical Laboratory, UK.

New English Bible (1970) Oxford University Press and Cambridge University Press, UK.

Owens, A.R. (1982) Digital signal conditioning and conversion. *Journal of Physics E: Scientific Instruments*, **15**, 789–805.

Pearce, J.R. (1983) *Scanning, A-to-D conversion and interference*, Technical report number 012/83, Solartron Instruments, Farnborough, UK.

Pearce, J.R. (1987) *What does your A/D converter do?* Colloquium on advanced A/D conversion techniques, IEE Digest No. 1987/48, pp. 1/1–14.

Pennington, D. (1965) *Piezoelectric Accelerometer Manual*, Endevco Corporation, Pasadena, CA.

Peyton, A.J. and Walsh, V. (1993) *Analog Electronics with Op-amps*, Cambridge University Press, UK.

Pleau, R. (1993) DT-Connect interface circumvents PC bottlenecks, *1993 Applications Handbook*, Vol. 2, No. 1, Data Translation, Inc., Marlboro, MA, pp. 138–139.

Pratt, W.J. (1974) Don't lean on a/d specs when you work with high encode rates. *Electronic Design*, **8** (12 April), 80–84.

Preston-Thomas, H. (1990) The International Temperature Scale of 1990. *Metrologia*, **27**, 3–10 and 107.

Prophet, G. (1983) Selecting Zener diodes for maximum stability. *Electronic Product Design* (March), pp. 59–60.

Quinn, T.J. (1990) *Temperature*, 2nd edn, Academic Press, London.

Rogers, J. (1995) Validity of calibration and test data: application of ISO/IEC Guide 25 (EN45001) or the ISO9000 series. *Engineering Science and Education*, **4**(5), 211–215.

Rudolf, F., van den Vlekkert, H. and Degrauwe, M. (1995) From airbags to printers. *Physics World*, **8**(11), 47–51.

Rusby, R.L. (1989) *Adoption of the International Temperature Scale of 1990*, ITS-90, National Physical Laboratory, Teddington, UK.

Scott, R.E. (1960) *Linear Circuits*, Addison-Wesley, Reading, MA.

Sheingold, D. (1986) *Analog–Digital Conversion Handbook*, Prentice-Hall, Englewood Cliffs, NJ.

Skinner, A.D. (1995) Calibration certificates: are you getting the certificate you need? *Engineering Science and Education*, **4**(5), 201–206.

Slater, J. (1994) Coded channels digital television by satellite, *IEE Review*, **40**(4), 175–179.

Smith, B.F. (1974) *High speed A/D and D/A converters, their effect on video signal fidelity*, Analog Devices, Norwood, MA.

Stever, S.D. (1989) An 8½-digit digital multimeter capable of 100 000 readings per second and two-source calibration. *Hewlett-Packard Journal*, **40**(2), 6–7.

Sydenham, P.H. (ed.) (1982) *Handbook of Measurement Science*, 1, Wiley, Chichester.

Sydenham, P.H. (ed.) (1983) *Handbook of Measurement Science*, 2, Wiley, Chichester.

Sydenham, P.H. Hancock, N.H. and Thorn, R. (1989) *Introduction to Measurement Science and Engineering*, Wiley, Chichester.

Taylor, H.R. (1974) *Economical data logger programs for the PDP-8*, Proceedings of Digital Equipment Computer User's Society, Zurich, p. 103.

Taylor, H.R. (1977) A comparison of methods for measuring the frequency response of mechanical structures with particular reference to machine tools. *Proceedings of the Institution of Mechanical Engineers*, **191**(16), 257–270; D49–D50.

Taylor, H.R. (1986) Teaching experiments with the dual ramp voltmeter. *International Journal of Electrical Engineering Education*, **23**(4), 311–320.

Taylor, H.R. and Rihawi, M.B. (1993) The dynamic thermometer: an instrument for fast measurements with platinum resistance thermometers. *Transactions of the Institute of Measurement and Control*, **15**(1), 11–18.

Taylor, T. (1995) Data acquisition: one step at a time. *Physics World*, **8**(10), 49–53.

Thompson, A.M. (1968) An absolute determination of resistance based on a calculable standard of capacitance. *Metrologica*, **4**(1), 107.

Tobey, G.E., Graeme, J.G. and Huelsman, L.P. (1971) *Operational Amplifiers; Design and Applications*, McGraw-Hill, Kogakusha, Tokyo.

von Hippel, A.R. (ed.) (1994) *Dielectric Materials and Applications*, Artech House, London.

Wilkie, T. (1983) Time to remeasure the metre. *New Scientist*, **45** (27 October), 258–263.

Wolfenbuttel, R.F. (1995) *Silicon Sensors and Circuits*, Chapman and Hall, London.

Wright, J. (1995) Technology is changing the way measurements are made. *Scientific Computing World*, (7), 35–38.

Young, C.A. (1987) *Error prediction techniques for enhanced high-accuracy analog-to-digital converters*, Colloquium on Advanced A/D Conversion Techniques, IEE Digest No. 1987/48.

Young, C.A. (1990) An enhanced method for characterising successive approximation converters, *IEEE Transactions on Instrumentation and Measurement*, **39**(2), 335–339.

Zaks, R. (1977) *Microprocessors*, 2nd edn, Sybex, Berkeley, CA.

Index